MARITIME RISK AND
ORGANIZATIONAL LEARNING

To Angeley

Maritime Risk and Organizational Learning

MICHAEL EKOW MANUEL
Accra, Ghana

CRC Press
Taylor & Francis Group
Boca Raton London New York

CRC Press is an imprint of the
Taylor & Francis Group, an **informa** business

CRC Press
Taylor & Francis Group
6000 Broken Sound Parkway NW, Suite 300
Boca Raton, FL 33487-2742

First issued in paperback 2017

Version Date: 20160226

ISBN 13: 978-1-138-07214-5 (pbk)
ISBN 13: 978-1-4094-1963-1 (hbk)

Visit the Taylor & Francis Web site at
http://www.taylorandfrancis.com

and the CRC Press Web site at
http://www.crcpress.com

Contents

List of Figures

List of Tables

Abbreviations

AB	Able Bodied Seaman (rating rank)
ABS	American Bureau of Shipping (American ship classification society)
ALARP	As Low As Reasonably Practicable
ANOVA	Analysis of Variance
APA	American Psychological Association
ASRS	Aviation Safety Reporting System
BBS	Behaviour-Based Safety
BRM	Bridge Resource Management
CCPS	Center for Chemical Process Safety
CCR	Cargo Control Room
CDI	Chemical Distributors Institute
CHIRP	Confidential Human factors Incident Reporting Programme
CRM	Crew Resource Management
CRM	Crew Resource Management; originally Cockpit Resource Management
DNV	Det Norske Veritas
DPA	Designated Person Ashore
DWT	Deadweight
EDH	Efficient Deck Hand (rating rank)
EEA	European Environmental Agency
EMCIP	European Marine Casualty Information Platform
FAO-EMPRES	Food and Agriculture Organization Emergency Prevention System
FOPM	Fleet Operating Procedures Manual
FRA	Federal Railroad Administration
FSI	Flag State Implementation
GISIS	Global Integrated Shipping Information System
GPR	General Purpose Rating
GT	Gross Tonnage
HERO	Hierarchical Epistemology and Realist Ontology (of risk)
HR	Human Resource
HSE	Health, Safety and Environment
HSEQ	Health, Safety, Environment and Quality
HSSE	Health, Safety, Security and Environment
IAEA	International Atomic Energy Agency
ICS	International Chamber of Shipping
IEC	International Electro-technical Commission

ILO	International Labour Organization
IMO	International Maritime Organization
INSJÖ	Informationssystemet om incidenter i sjöfarten (*database of incidents in shipping – Swedish*)
INTERTANKO	International Association of Independent Tanker Owners
ISM	International Safety Management (Code)
ISO	International Organization for Standardization
IT	Information Technology
ITOSF	Informal Tanker Operators Safety Forum
KPI	Key Performance Indicators
LI	Leader Inclusiveness
LNG	Liquefied Natural Gas
LTI	Lost Time Injuries
LTIF	Lost Time Injury Frequency
MARS	Marine Accident Reporting Scheme
MET	Maritime Education and Training
METI	Maritime Education and Training Institution(s)
MLC	Maritime Labour Convention
MRM	Maritime Resource Management
M-SCAT	Marine – Systematic Cause Analysis Technique
NASA	National Aeronautics and Space Administration
NK	Class Nippon Kaiji Kyokai
NTSB	National Transportation Safety Board
OCIMF	Oil Companies International Marine Forum
OECD	Organisation for Economic Co-operation and Development
OHS	Occupational Health and Safety
OL	Organizational Learning
P&I	Provident and Indemnity
PEC	Protection and Environmental Committee
PPE	Personal Protective Equipment
PSSR	Personal Safety and Social Responsibility
QDA	Qualitative Data Analysis
SAFIR	**Saf**ety and **I**mprovement **R**eporting System / **Saf**ety **I**nformation **R**eporting
SARF	Social Amplification of Risk Framework
SBMA	Subic Bay Metropolitan Authority
SIRE	Ship Inspection Report Programme
SMS	Safety Management System
SOLAS	International Convention for the Safety of Life at Sea
SQEMS	Safety, Quality and Environmental Protection Management System
SQMM	Safety and Quality Management Manual
SRA	Society for Risk Analysis
STAR IPS	Star Information & Planning System

STCW Code	Seafarers' Training, Certification and Watchkeeping Code
STCW Convention	International Convention on Standards of Training, Certification and Watchkeeping for Seafarers 1978, as amended in 1995 and 1997
TMSA	Tanker Management and Self Assessment
TPS	Team Psychological Safety
TSG	Tankers Supervisory Group
UMS	Unmanned Machinery Space
USCG	United States Coast Guard
VLCC	Very Large Crude Carrier
WE	Worker Engagement

Preface

The journey that has resulted in this book was motivated by a number of factors. I worked as a ship officer on seagoing ships for 11 years, culminating in an appointment as master. During this period I often observed how the best-laid safety mechanisms, plans and procedures were undermined by social dynamics on board ship and between ship and shore. Since then I have lectured in a traditional maritime education and training institution and been exposed to an international regime of maritime education and training in various national contexts. I have found this regime to focus on competence, skill and the perfecting of professional practices which have their origins in centuries of commercial shipping. This focus limits the engagement of industry with socio-psychological issues (the human factors) especially at the educational level. The gravity of these issues in industry, as I perceived them, and the paucity of educational attempts to address them, gives cause for concern. In the industry as a whole and over decades, there has been an increasing emphasis on the need for an enhanced safety focus and an accompanying increase in legislation (with the associated administrative burden) to achieve those ends. There has also been, on the sidelines of risk management, a creeping appreciation in the literature of the role of sociological factors in risk management in the maritime industry. The influence of such factors are significant to the achievement (or otherwise) of safety goals. However, because such social dynamics in construing risk – and in what we do with risk information – are acknowledged to be difficult to measure, legislate on, or audit, they are often left to the periphery of risk cognition. If behaviour – individual and organizational[1] – are so influenced by perceptions, for example, then to ignore the processes that inform these perceptions and simply legislate on what is obvious, measurable and auditable, leaves the industry handicapped in its ability to address risk sufficiently.

Significant work has been done in the world of risk research. It has, appropriately, been focused on high-risk industries. However, while the maritime industry is included in this category, there has been a relative lack of industry-specific research focusing on the construal of risk and how it is impacted by the engagement of the worker, team dynamics and social learning. While the value of these constructs has not been disputed, they are mainly supported only by anecdotal evidence. This book has been motivated by a desire to bridge this gap by inquiring into the processes that inform organizational learning in shipping

---

1 The words 'organization' and 'organizational' are spelt as shown instead of 'organisation' or 'organisational' respectively (although the text of this work is based on UK English spelling). This is because of the dominance of the former couple of spellings in the literature and in International Maritime Organization (IMO) texts.

companies in particular and in the industry in general. The maritime industry, like other high-risk industries, has long been recognized as one exposed to peril and risk of different levels. However, it is unique in many of its characteristics, particularly those associated with social settings on board ship and also the global influences at play.

In recognition of the risk inherent in the industry, many stakeholders – including international organizations, national and private interests – have worked to mitigate the adverse impacts of risk in industry. These efforts have yielded much fruit but have mainly been in the form of legislation and the intuitive apprehension of risk and its construal. Research into social influences on risk, and its construal and management, has mainly been limited to the teams on board ship and has neither explicitly sought to clarify the 'team links' between ship and shore, nor to examine the contexts of the shore-based company. This work has sought to investigate these processes, as well as to explore the input of seafarer risk epistemology into the organizational learning practices of shipping organizations and the resulting potential amplification and/or attenuation of risk signals.

The maritime industry's focus on risk avoidance has tended to evolve in two directions. One is the focus that ends in blame, with attendant punitive sanctions as deterrents and the enactment of regulations to optimize safety behaviour. On the other hand, there has been a focus by others on learning. This latter approach attempts to limit the presence of blame and the adverse effect it is believed to have on safety optimization and learning. What has been missing, perhaps, is research that shows how this plays out in the industry. How do shipping companies learn? Have the relatively contemporary concepts of 'organizational learning' and 'learning organizations' gained any traction in shipping or are they remote theoretical concepts that are seldom/rarely applied to the safety world of shipping? The opinion is often expressed that there is learning from accidents. However, with the observation of repetition of accidents with similar dynamics comes the question: do shipping companies really learn and if they do, how is this learning achieved in contemporary times and what kind of learning is dominant? Among other things, this is what has motivated this work, a work which is premised on the philosophical tenets of risk and socio-psychology.

Teichmann and Evans (1999, p. 1) define philosophy as the 'study of problems which are ultimate, abstract and very general. These problems are concerned with the nature of existence, knowledge, morality, reason and human purpose'. Quinton's elaborate view is particularly noteworthy.

Philosophy is rationally critical thinking, of a more or less systematic kind about the general nature of the world (metaphysics or theory of existence), the justification of belief (epistemology or theory of knowledge), and the conduct of life (ethics or theory of value). Each of the three elements in this list has a non-philosophical counterpart, from which it is distinguished by its explicitly rational and critical way of proceeding and by its systematic nature. Everyone has some general conception of the nature of the world in which they live and of their place in it. Metaphysics replaces the unargued assumptions embodied in such a conception with a rational and organized body of beliefs about the world as a whole. Everyone has occasion to doubt and question beliefs, their own or those of others, with more or less success and without any theory of what they are doing. Epistemology seeks by argument to make explicit the rules of correct belief formation. Everyone governs their conduct by directing it to desired or valued ends. Ethics, or moral philosophy, in its most inclusive sense, seeks to articulate, in rationally systematic form, the rules or principles involved (Quinton, 1995, p. 666).

To these 'philosophical' ends then, this study attempts to give some empiric evidence of concepts and views that – though perhaps intuitively and anecdotally existing – have little coverage in the research literature with the particular links established here (between risk, learning, engagement and accident philosophy) and especially specific to the maritime context.

In contributing to the discussion, this study shows that there is merit in complementing the valuable work of legislation with enhanced education (imbuing of social norms) and markets as a means of gaining an appreciation of, an emphasis on and benefits from the socio-psychological parts of risk, its construal and its management.

The text of this study includes numerous references to and citations of individuals and entities. Such citations or references do not necessarily imply unqualified agreement or endorsement of the views (whole or in part) represented by the cited individuals/entities. They are given only to the extent that they contribute (either in agreement or in disagreement) to the discussion or point being made in the specific context in which they are used.

Michael Ekow Manuel
June 2011

Acknowledgements

I am indebted to many who have sacrificed for me and supported me during the studies that led to this book. My family sacrificed the most in letting me finish this work. To you, Angeley, Elsie and Joel, my sincerest thanks.

Sincere gratitude to my parents, brothers and sisters (biological and in-law) and to dear friends in Sweden for the company and encouragement.

My gratitude is also due to the World Maritime University for the grant of the Sheldon Kinney Fellowship; special thanks to Professors Jens-Uwe Schröder, Proshanto Mukherjee, Takeshi Nakazawa, Patrick Donner and Rajendra Prasad for the help in contacting different institutions and for open doors for discussions; to all faculty and staff of the university – particularly the library and IT staff and faculty assistants – for friendly, professional and superlative assistance; to the Regional Maritime University in Ghana; to colleagues Kevin Ghirxi, Abhinayan Basu, Kofi Mbiah, Rajendra Prasad and Gabriel Figueiredo for company on the journey; to my external consulting group – seven academics and industry experts who gave me insightful and very helpful feedback; to Andreas Bach (and pilots of Malmö Port) and Richmond Quayson (and pilots of Tema Port); to the six shipping companies who welcomed me and freely shared with me valuable time, information and access to software and documentation; to WMU Classes of 2007, 2008 and 2009 for focus group discussions and friendships; to all the people at Ashgate Publishing, especially Margaret, Guy and Jude, for excellence in what they do and to the many others who helped in various ways.

The study that informed this book has exposed me to a lot of human effort to 'control' the future. It was inevitable that it would also highlight the inadequacy of humanity and our often feeble attempts to grapple with uncertainty. To my mind it ultimately points to our need to acknowledge dimensions and divinity transcending our limited apprehension of even the history we have lived, not to mention the present or the future. To Almighty God, whom I serve with all my heart, and who is the supreme guarantee against the ultimate risk, I reserve the greatest thanks and the greatest praise, through Jesus Christ.

Acknowledgements for Figures

1.1   Role of people in recovery process

Source: Kanse, L. (2004). *Recovery uncovered: How people in the chemical process industry recover from failures.* Unpublished PhD dissertation, Technische Universiteit Eindhoven, Eindhoven, Netherlands. Used with permission.

1.2   Accident frequency trends in maritime industry

Source: Det Norske Veritas (2009). Used with permission.

2.1   The Social Amplification of Risk Framework (SARF)

Source: Kasperson, J.X., Kasperson, R.E., Pidgeon, N. and Slovic, P. (2003). The social amplification of risk: Assessing fifteen years of research and theory. In N. Pidgeon, R.E. Kasperson and P. Slovic (eds.), *The social amplification of risk* (pp. 13–46). Cambridge: Cambridge University Press. Used under the permission terms of the Cambridge University Press for single figures or tables.

3.2   Constructs and processes associated with organizational learning

Reprinted by permission, Huber, G.P., Organizational learning: The contributing processes and the literatures, *Organization Science*, 2(1), 88–115, 1991, Copyright © The Institute for Operations Research and the Management Sciences (INFORMS), 7240 Parkway Drive, Suite 300, Hanover, MD 21076 USA.

3.3   A taxonomy of human motivation

Reprinted from *Contemporary Educational Psychology* 25, Ryan, R.M. and Deci, E. L. Intrinsic and extrinsic motivations: classic definitions and new directions, 54–67. Copyright © (2000) Academy Press.

3.4   Worker engagement activities

Reproduced under the terms of the Click-Use Licence of the UK Office of Public Sector Information (OPSI).from Health and Safety Commission. (2006). Improving worker involvement – Improving health and safety (No. CD207 C10 04/06). Suffolk: Health and Safety.

3.5   Approaches to worker engagement

Source: Adapted from Cameron, I., Hare, B., Duff, R. and Maloney, B. (2006). An investigation of approaches to worker engagement. Norwich: HSE Books, under the terms of the Click-Use Licence of the UK Office of Public Sector Information (OPSI).

7.2   Problem solution model

Source: Kim, D.H. (1993). A framework and methodology for linking individual and organizational learning: Applications in TQM and product development. Unpublished PhD thesis, Sloan School of Management: Massachusetts Institute of Technology. Used with permission.

7.3   Problem articulation model

Source: Kim, D.H. (1993). A framework and methodology for linking individual and organizational learning: Applications in TQM and product development. Unpublished PhD thesis, Sloan School of Management: Massachusetts Institute of Technology. Used with permission.

7.5   Conceptual model of Worker Engagement

Reproduced under the terms of the Click-Use Licence of the UK Office of Public Sector Information (OPSI) from Cameron, I., Hare, B., Duff, R. and Maloney, B. (2006). An investigation of approaches to worker engagement. Norwich: HSE Books.

Acknowledgments for Tables

2.2   **A systematic classification of risk perspectives/approaches**

Source: Ortwin Renn. Concepts of risk: A classification. In Sheldon Krimsky and Dominic Golding (eds.), *Social theories of risk* (pp. 53–79). Copyright © (1992). Reproduced with permission of ABC-CLIO, LLC.

Chapter 1

Introduction

It is unwise to be too sure of one's own wisdom. It is healthy to be reminded that the strongest might weaken and the wisest might err.

Mahatma Gandhi[1]

Purpose and Outline

This book focuses on how shipping companies learn from risk perceptions and how that learning influences the companies' work culture, group dynamics and individual motivation. As an introduction to the questions and analyses that is the book's focus, Chapter 1 presents a number of examples of accidents in high-risk industries taken from official accident report extracts and the safety literature and is intended to highlight a common and recurring thread in contributory factors to accident causation. It also gives a brief outline of the shipping industry and presents the book's aims and potential value.

Social Dynamics in Accident Causation: Examples

One of the most significant current issues in high-risk industries is the role and influence of human factors in operational efficiency, safety and environmental protection. Recent developments in the supply–demand dynamics of manning in the maritime industry have resulted in increasing interest in and acknowledgement of the need to address human factor issues, not only at an individual level, but also at team and organizational levels. A number of accidents, incidents and near-misses over decades have increasingly highlighted the fact that the most competent of operators and systems are fallible and that the augmentation of knowledge by drawing on aggregate team knowledge and perceptions of risk makes risk management better. Following are narratives of some accidents that serve to illustrate some of these dynamics.

The Kegworth Accident

On the 8 January 1989, British Midland Flight 92 en route from Heathrow (London) to Belfast, while trying an emergency landing at the East Midland Airport, crashed at the edge of the M1 motorway near Kegworth, Leicestershire. One engine (of

---

1   Retrieved 4 May 2009 from http://thinkexist.com/quotation/it_is_unwise_to_be_too_sure_of_one-s_own_wisdom/11448.html.

the two-engine plane) had caught fire during the flight and in an attempt to control the emergency situation the pilots shut down the wrong engine (the right engine instead of the left engine). Many factors were identified to have contributed to the accident (see the 152-page report by the Air Accidents Investigation Branch of the UK, 1990 for a detailed analysis). Despite all the technical, training and ergonomic issues that were identified as proximate or distant causal factors, this accident could have been avoided had the system and role-status perceptions being such that cabin crew would have informed the flight deck that they had seen fire in the left engine. In the words of the report:

> It is extremely unfortunate that the information evident to many of the passengers of fire associated with the left engine did not find its way to the flight deck even though, when the commander made his cabin address broadcast, he stated that he had shut down the 'right' engine. The factor of the role commonly adopted by passengers probably influenced this lack of communication. Lay passengers generally accept that the pilot is provided with full information on the state of the aircraft and they will regard it as unlikely that they have much to contribute to his knowledge ... The same information was available to the 3 cabin crew in the rear of the aircraft, but they, like the passengers, would have had no reason to suppose that the evidence of malfunction they saw on the left engine was not equally apparent to the flight crew from the engine instruments ... In addition cabin crew are generally aware that any intrusion into the flight deck during busy phases of flight may be distracting ... However it must be stated that had some initiative been taken by one or more of the cabin crew who had seen the distress of the left engine, this accident could have been prevented. (Air Accidents Investigation Branch of the UK, 1990, p. 106. Reproduced under the terms of the Click-Use Licence of the UK Office of Public Sector Information (OPSI)).

The report further noted (p. 109) that 'it could be argued that the pilots of this aircraft did not make effective use of the cabin crew as an additional source of information'. While this may have been understandable in light of the whole accident report findings, it is nevertheless a sad omission, the avoidance of which could have saved many lives.

Herald of Free Enterprise

The *Herald of Free Enterprise* accident occurred, on the 6 March 1987, when this ferry sailed from the port of Zeebrugge with a bow door open leading to the significant ingress of water and subsequent capsizing of the ship; 150 passengers and 38 crew members lost their lives. The proximate cause of the accident – the non-closure of the bow door by the designated crew member – was attributed to 'errors of omission on the part of the Master, Chief Officer and the assistant bosun' (UK Department of Transport, 1987, p. 14). However, the investigation report also noted more fundamental but relatively latent issues, observing that, 'from top to bottom the body corporate was infected with the disease of sloppiness' (1987, p. 14). This accident shocked the world of shipping. It led to a rash of new legislation (Lavery, 1989) and an increased focus on the shipping company's role in accident causation and prevention. It also highlighted how the reticence of masters/ship crew to challenge shore management when put under inappropriate commercial pressure, compromises or jeopardizes the safety of the ship and crew. This is indicative of the way risk information based on individual perceptions is attenuated by organizational systems, culture and economic priorities.

The report indicates (p. 23) that a few years earlier (in October, 1983), a sister ship of the *Herald of Free Enterprise*, the *Pride of Free Enterprise*, had sailed from Dover with all doors open because an assistant bosun had fallen asleep just as in the case of the *Herald of Free Enterprise*. The organizational structures of the company had been unable to achieve proactive learning from this earlier incident and indeed from many other communications of risk from ship to shore.

While the subsequent focus on organizational systems have been significant, the underlying social dynamics, onboard and between shore and ship, that lead to reticence concerning the communication of risk and the lack of fundamental proactive learning seem not to have been sufficiently addressed in the maritime industry.

Paediatric Death

Accidents and even fatalities that occur due to sub-optimal social dynamics in teams and the resulting attenuation of risk information and signals are not restricted to the aviation and maritime industries. One other high-risk domain where this is an issue is the medical field. According to Helmreich (2000, p. 784), an eight-year-old boy died unnecessarily, due to the actions (or inactions) of an anaesthetist during elective surgery on the eardrum. Among other arguably very unprofessional conduct, the attending anaesthetist was observed to be dozing during a critical time in the surgery. Some nurses observed this but did not speak up because they 'were afraid of a confrontation.' The young boy subsequently died, despite the efforts of an emergency team that had been summoned. The causes of death were attributed in no small way to the actions (and inactions) of the anaesthetist.

This is not an isolated case. Lingard et al. (2004, p. 330) cite the Chief Coroner of the Province of Ontario in Canada as reporting 'communication difficulties at all levels [emphasis added] of the hospital, including doctors to doctors, doctors to nurses, nurses to nurses and nurses to doctors, as the primary cause of errors leading the death of a paediatric patient'.

Green Lily

In November, 1997, the reefer vessel *Green Lily* grounded off the Shetland Islands, after having sailed into severe weather. One life (of a rescue helicopter winchman) was lost with the ship being driven ashore resulting in a total loss. The accident report indicates that:

> The master received no external pressure to sail. He was aware that the vessel would be heading into adverse weather and that progress would be slow. He was also aware that adverse weather was forecast for several days ahead, and that if he chose not to sail, the vessel would be significantly delayed. When sufficiently clear of the land, he intended to turn the vessel on to a more southerly heading to reduce the adverse effect of the wind on the vessel's speed. In deciding to sail on 18 November, the master was optimistic that the prevailing and predicted weather conditions outside Lerwick would not unduly hinder the vessel's progress. He should have considered the worst predicted conditions and their effect. Although at least one officer was concerned about the master's decision to sail, no one openly questioned him. After clearing Bressay, the vessel was effectively hove to in south-east force 9 winds. The master recognised that the weather conditions were worse than he had expected and that progress would be much slower than he had hoped. He had the opportunity of returning to Lerwick but chose not to do so, in the hope that the weather would improve. Having decided to sail, his decision not to return to harbour was possibly influenced by his not wishing to be seen as having failed to consider the worst predicted conditions. The reluctance of anyone on board to question the master's decision to sail from Lerwick, and his decision not to turn back after realising he had failed to consider the worst predicted weather conditions, suggests an autocratic style of management. A less authoritarian style might have encouraged greater discussion of the issues and would have enabled decision-making shortcomings to be identified at the outset. (Marine Accident Investigation Branch, 1999. Reproduced under the terms of the Click-Use Licence of the UK Office of Public Sector Information (OPSI)).

It is pertinent to ask why the master sailed in the absence of any external pressure to do so. One plausible reason could be that 'pressure to sail' results, not only from overt statements from the shore office, but also from the tacit prioritization of economics over safety or even the lack of significant and overt statements that create 'pressure not to sail' where there is doubt. This kind of latent organizational culture may also have influenced the crew in their unwillingness to state their perceptions of risk.

Bow Eagle

On the 26 August 2002 the fishing vessel *Cistude* collided with the Norwegian-registered chemical tanker *Bow Eagle*. The collision resulted in the loss of 4 lives (from *Cistude*) and the spilling of 200 tonnes of ethyl acetate (from *Bow Eagle*). After the collision, the *Bow Eagle* failed to stop to render assistance. The events in the immediate aftermath of the accident as indicated in the accident report are of particular interest in this book's context. The following[2] is an extract from the original report:

> After hearing the PAN PAN PAN message [from the damaged *Cistude*], at about 03H45, the lookout [of the *Bow Eagle*] asked the officer to inform the master. The officer did not follow this suggestion. Instead he demanded that the lookout keep the incident a secret – not to speak about it – which the lookout obeyed. This exchange took place in Tagalog (Philippines vernacular). The officer reiterated this the next day with the same results. At 04H00, the watch was handed over to the second officer without any special comments. (Bureau Enquêtes – Accidents/Mer [BEAmer], 2003, p. 51)

Despite the obvious potential presence of latent causative factors and a real opportunity to query contributory factors to risk situations (for example, social relations, attenuation of risk information), this only subsequently (at least at the level of investigating for lessons learned) led to the jailing of the officer on watch on the *Bow Eagle* ('Officer of Norwegian tanker jailed over collision,' 2003). This seems to be a typical reaction in the maritime industry.

Bow Mariner

The *Bow Mariner* accident is perhaps the one accident that most clearly shows the potential effects of failure in organizational learning in the maritime industry as well as the attenuation (or complete inhibition) of risk signals/perceptions in the risk cognition process.

On Saturday 28 February 2004 the chemical tanker *Bow Mariner*, while engaged in tank-cleaning, caught fire, exploded and sank about 45 nautical miles east of Virginia, USA. The accident resulted in the death of 3 crew members and 18 missing (presumed dead) along with a substantial release of a cargo of ethyl alcohol and fuel. According to the report 'contributing to this casualty was the failure of the *operator*,[3] and the senior officers of the *Bow Mariner*, to properly implement the company and vessel Safety', Quality and Environmental Protection Management System (SQEMS) (United States Coast Guard [USCG], 2005, p. 1). On pages 42 and 43, the report notes that:

> The master, chief officer and chief engineer of the *Bow Mariner* were Greek and the remaining officers and crew were Filipino. The authority and responsibility of the senior shipboard management is spelled out in 6.1.3 of the SQMM [Safety and Quality Management Manual], as well as Sections 2.2 (master), 2.3 (chief

---

2   Translation from the original French by author.

3   All references to specific individual and company names, though present in this particular report, have been removed from this text and italicized text used for substitute references. The name of the ship is real.

officer) and 2.4 (chief engineer) of the FOPM [Fleet Operating Procedures Manual]. Under the SQEMS [Safety, Quality and Environmental Management System], the master has 'total responsibility' for the operation, seaworthiness and safety of the vessel at all times. The chief officer is also designated as the safety officer, responsible for maintenance of equipment and training of personnel, in addition to his other duties. All three senior officers are charged with implementing the SQEMS.

Section 2.1.1 of the FOPM describes the master's authority as follows:

> The master has full authority over all persons (personnel and passengers) onboard his vessel. The Master's authority is not questioned and must be supported and maintained by onboard personnel. Orders must be carried out and obeyed as said, in letter and in spirit. Refusal to do so is grounds for prompt disciplinary action, including possible termination of employment.

Such absolute authority is not unusual aboard seagoing vessels. Indeed, many would argue such absolute authority is essential to maintaining good order and discipline. But on the *Bow Mariner* the distinctions between the Greek senior officers and Filipino crew were remarkable. Filipino officers did not take their meals in the officer's mess, were given almost no responsibility and were closely supervised in every task. The second assistant engineer, who was working aboard a vessel *managed by this company* for the first time, was upset when he was chastised on his first day aboard because he inquired about his management and administrative duties. The chief engineer sternly told him that he would be given verbal job orders daily, was to do only as he was told and would have no administrative duties beyond making log entries. In contrast, Section 2.4.2 of the FOPM spells out significant duties for the second engineer – duties the chief engineer on the *Bow Mariner* was not prepared to entrust to his subordinate officers. This contrast between the content of the SQEMS and actual practice aboard the *Bow Mariner* was pervasive. The lack of trust was apparent on deck as well. The surviving deck crew reported that the chief officer would not sleep, beyond short naps in a chair in the CCR [Cargo Control Room], during cargo operations. They stated this practice was common aboard vessels *managed by this company*. The chief officer performed all management and administrative duties himself, including the preparation of plans for cargo loading/unloading, ballast management, tank cleaning and gas freeing, training and drills. He did not delegate or attempt to train the junior officers to perform any of these tasks, either to reduce his own workload or provide for their professional growth. As a result the Filipino crew had little knowledge of the technical aspects of their job, so much so that they failed to question unsafe actions or procedures. When questioned about what they would do if instructed to do something unsafe by one of the senior officers, each crewman replied that they would do as they were ordered. (One crewman said that the orders of the Greeks were 'like words from God'.)

This lack of technical knowledge and fear of the senior officers explains why the crew did not question the master's unsafe order to open all of the empty tanks; they either did not know about the danger or were not inclined to question the master's order. The fear of the Greek officers extended also to the galley. *One messman* reported that he did not like the Greek officers because they were verbally abusive to him and constantly threatened to send him home if he did not work harder or faster. *The chief cook* was likewise afraid of losing his job. While these may have been the usual complaints of the lowest ranking crewmen aboard ship, there can be no question that such fear can lead to a shipboard culture where safety takes a backseat to preserving one's livelihood. This attitude toward Filipino officers and crew was not limited to the *Bow Mariner*. As part of this investigation, the investigating officer visited a sister vessel, the *Bow Transporter*, in Singapore. During that visit many of the same attitudes were observed. The Filipinos were only permitted to speak to the investigating officer and Singapore officials in the presence of the senior officers, leading to obvious nervousness. Nevertheless, several crewmembers made statements confirming the same cultural divide existed aboard the *Bow Transporter*. The chief officer reported that he planned all of the cargo and tank cleaning operations, and also remained awake throughout all cargo operations. And engineers reported that they were not permitted to test the IGS [Inert Gas System] in the absence of the chief engineer, a task any licensed officer on a tank ship should be able to perform. Probably the most telling evidence of the lack of cohesiveness in the crew of the *Bow Mariner* was their response to the explosion. Although the official language of the crew was English, the *captain* and *the chief engineer* were conversing in Greek when they assembled with the crew aft of the accommodations. *The messman*, who was with this group, reported that he and the other crewmen were simply waiting for someone to tell them what to do. Those instructions never came. The final blow came when *the captain* ignored questions from *the third officer* about whether a distress signal had been sent. Instead of an organized, thoughtful response, the situation deteriorated to 'every man for himself.' *Company* officials have defended the *captain's* actions and the crew's reaction after the explosion, citing emotional trauma triggered by the explosions, fire and immediate list.

However, such trauma is expected and is precisely the reason that crews must be thoroughly trained and frequently drilled – so that they will react instinctively in an emergency just as they have been trained. The 'trauma explanation' is also suspect given that far less experienced crewmembers controlled their emotions and reacted professionally. *The captain* abandoned ship without sending a distress signal or conducting a muster, and left behind crewmembers he knew to be alive. Such conduct reflects his failure to conduct regular, realistic drills to prevent just such a reaction (United States Coast Guard [USCG], 2005, pp. 42, 43. Used under the terms of the USCG).

A couple of points are noteworthy:

- The USCG found the 'actions of *the third officer*, making his first trip as a licensed officer commendable' and 'recognized him for his heroic efforts' which saved his life and that of 5 others (United States Coast Guard [USCG], 2005, p. 46).
- The *Bow Mariner* was owned by the same company that owned the *Bow Eagle*. *Bow Mariner* was operated by *a company*, which had previously had a relatively safe and successful relationship with *the owning company* operating 18 ships on their behalf.[4]

Boiler Rupture on the Cruise Ship Norway

On 25 May 2003, the S/S *Norway* had an accident involving the rupture of a boiler while berthed in Miami, Florida. The accident resulted in 8 crewmembers sustaining fatal injuries, and 17 others suffering injuries ranging from serious to minor (National Transportation Safety Board [NTSB], 2007). The prima facie causes of the accident according to the report were:

- The lack of adherence to water chemistry composition limits and procedures by both the water chemistry subcontractors and *shipowners* during wet lay-up periods, leading to pitting from oxygen corrosion.
- Failure to take number of boiler cycles into account during maintenance.
- Severe thermal transients from heating and cooling the boilers too quickly and from constraints created by frozen boiler support feet.
- Use of questionable weld repair procedures.
- Lack of appropriate non-destructive testing by the *classification society* surveyors and *shipowner* inspectors to determine whether cracks were present.
- Inadequate survey guidance from *the classification society* to its surveyors.
- Failure to repair cracks into which copper had been inappropriately introduced (p. 42).

This report evidently attributed immediate accident causation to technical reasons. However, it also highlighted significant latent causative factors. Over a period of at least two decades, the owners and operators of the ship had been repeatedly warned about lapses in the operational routine and maintenance of the boiler of the ship *Norway* (pp. 14, 21–23). It showed how management did or did not value the risk perceptions of operators, from ignoring warnings, to acknowledging warnings, to making verbal/written commitments for action, without following through.

*****

---

4   Information from owning company annual report, 2005.

It is worth noting, that in the maritime accidents, the ships involved had passed various levels of audits and surveys and had valid certificates indicating their seaworthiness according to the prevailing standards.

Common threads run through all of these accidents and many others like them; there are inhibitory factors that limit *interpersonal risk-taking*[5] in teams in high-risk industries and that put pressure – tacit or otherwise – on ship crew to cut down on safety margins (Thai and Grewal, 2006). These factors, which may be subsumed into the larger context of the failure of organizational learning, work against team members, especially subordinates, contributing potentially relevant risk information. Individuals are not prepared to take the risk of 'challenging' superiors (whether ship or shore based), questioning courses of action, or even noting concerns. Perceived superior competence of other team members, role-based status in teams and perceived insignificance of the information one holds, are some of the reasons for the existence of social dynamics that make team members feel insecure. These perceptions may be exacerbated by a culture that limits learning at a deeper than superficial level. In each of the accidents described, it is easy to put the blame on what has been called 'the sharp end' and find 'bad apples' there; an incompetent and sleeping anaesthetist, erring engineers, self-opinionated ship officers, a hasty and/or autocratic master or an irresponsible assistant bosun. However, in every one of the accidents, latent and arguably more dangerous factors are evident. These are more dangerous because they are not as simple as may initially appear and also lead to recurrence even after the bad apple is 'eliminated', designed out of the system, replaced by technology or subjected to more procedures. As Dekker (2006, p. 17) puts it, 'accidents are actually structural by-products of a system's normal functioning' and to quickly and self-righteously apportion blame to the sharp-end limits learning. The better way is to treat 'human error' as a symptom of deeper malaise and to try to understand why people did what they did given the circumstances and contexts in which they were operating. This implies that those people must feel, to a degree, a sense of responsibility for their actions but that this sense should not be interpreted as blame-worthiness that will lead to a lack of speaking up and interpersonal risk-taking. Operators being able to speak up and contribute to fundamental learning in organizations is an imperative because 'people are vital to creating safety ... and are the only ones who can negotiate between safety and other pressures in actual operating conditions' (Dekker, 2006, p. 16). Interpersonal risk-taking and its input into organizational learning and safety is thus essential to teamwork in all high-risk industries. However, it may be postulated that the social dynamics and uniqueness of the maritime industry make it even more critical that these issues be examined and yet the paucity of such examination in the industry as compared to others

---

5   Interpersonal risk-taking is the ability of members of a team to take the risk of questioning the actions of others in the team and making their own perceptions of risk known.

is apparent. A limited list of factors which make the shipping industry unique includes:

- Multicultural crewing.
- Distinct differences in team structure, for example, in rank, experience, age and nationality.
- The traditionally hierarchical nature of shipboard teams, blending civil structures with quasi-military norms and an emphasis on competence.
- The gender bias towards masculinity.
- The multitasking nature of operations.
- The potentially harsh physical environment in which operations are carried out and remoteness from other teams.
- Current trends in manning levels.
- High labour mobility due to contractual nature of employment.
- Limited social options acting as stress alleviators exacerbated by extensive periods of contracts during which crew are away from significant shore-based social contact.
- Limited selection procedures for team compatibility as well as limited opportunities for team-based training.
- The not uncommon practice of salary differentiation based on nationality for the same rank/work.

These factors, among others, make the shipping setting unique enough to warrant a critical analysis of the probable lack of, or disruption of, learning and the place of seafarer input into risk management specific to this industry. Previously, academic and industry response to such issues has rightly been to ask for better training in non-technical skills for operators 'at the sharp end' (Flin, O'Connor and Crichton, 2008). As noted by Lingard et al. (2004, p. 330), teamwork is often treated with respect to a cluster of factors/behaviours for example, 'leadership, technical skills, coordination, situational awareness, communication' and the production of 'multidimensional schemes to capture the quality of teamwork'. This has led to the requirement in many transport industries for Crew Resource Management (CRM) courses (Barnett, Gatfield and Pekcan, 2003; Federal Railroad Administration (FRA), 2006; Helmreich Merritt and Wilhelm, 1999). While this is laudable and can be said to have significantly improved the situation, it is felt that specific organizational factors, that should foster a maritime industrial paradigm in which interpersonal risk-taking and organizational learning can prosper, has not been examined in theoretical detail especially as relates to shore-based input. The evolution of CRM in aviation (originally Cockpit Resource Management) has seen the involvement of 'the entire flight crew, air traffic controllers and aircraft maintenance personnel' (FRA 2006, p. 1), a holistic approach that is arguably absent in the maritime industry. From an educational and management point of view, it is relatively easy to focus on the bridge team (or even the ship team), whether it is with respect to technical skills or team training. However, it is also necessary that attention be given to ship–shore teamwork and the way in which organizations create or inhibit a communicating and learning culture. In other words, accident causation is not only about people (as individuals or teams) or the technology they work with alone. Significantly, it is also about places (the organizational systems and associated cultures) in which they work. 'Places' here refers to, not only physical locations, but also to social settings of interaction.

The multiple-factor causality of accidents and deviations from quality is well-documented (Hollnagel, 2004; Reason, 1990, p. 197; Wagenaar and Groeneweg, 1987). While acknowledging this, the current study focuses on those factors that impinge on the inability of the operators to act as a team, the lack of 'confidence' of subordinates in contributing safety-critical information to other organizational members, and the non-existence of a learning culture where this is required.

One key consideration in accident prevention is to avoid the creation and stimulation of conditions that facilitate accident causation mechanisms irrespective of the source of these mechanisms. In the not uncommon case that such attempts are unsuccessful, consequences of undesirable system conditions can be mitigated by appropriate actions (and avoidance of certain actions) in a recovery mode (Kanse, 2004; van der Schaaf, Frese and Heimbeck, 1996). Figure 1.1 shows this process of recovery and the important role of people in 'detecting, diagnosing and correcting failures, ranging from human errors to technical or system faults' (Kanse, 2004, p. 3).

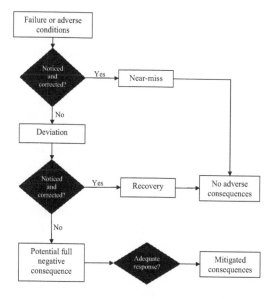

Figure 1.1   Role of people in recovery process
Source: Adapted from Kanse, L. (2004)

Whether one looks at the originating 'conditions' and associated outcomes as 'errors/failures' or as 'performance variability' of the human element or of whole organizational systems, and irrespective of the accident perspective one takes, the importance of intervention is real. In acknowledging that the systems that humans create are 'not basically safe' (Dekker, 2006, p. 16), it is worthwhile examining in greater detail and with increased specificity for the maritime industry, the role that humans in teams – adequately supported by organizational systems and features – can play in noting deviations or entropic drifts towards the margins of safety. The potential of the latter 'migration toward the boundary' (Rasmussen, 1997, p. 189) is best appreciated by people communicating at all levels, who can then mitigate the consequences of actions/inactions initiated by themselves, others or the system as a whole. It is therefore important that when features of a safety system (especially latent ones) are noticed, they are brought to the attention of 'system controllers' in an uninhibited way. The accident 'sequence' is characterized by a latent phase which has been termed the incubation stage (Johnson, 2003; Turner and Pidgeon, 1997), metastable stage (Hale and Glendon, 1987, p. 14) or metaphorically as a period when 'resident pathogens' (Reason, 1990, p. 197) are developing. This latency is inherent in almost all high-risk industries almost as a by-product of 'normal work' and may lead to 'normal accidents' as a result of tight coupling and high interactive complexity in socio-technical systems (Perrow, 1999). The very fact of this latency and the uncertainty and ambiguity associated with it, means that the appropriate intervention is only possible when insight and foresight are built into such systems as a step towards anticipation and resilience. This presupposes that individuals at all levels, who may have such insights and foresights and who may be first to notice symptoms of the conditions that could potentially lead to failure, are able to speak up, thus contributing to the building up of organizational resilience. The ability to communicate such individual perceptions of risk is critical to any effort to recognize or recover from adverse latent condition and/or deviations from system safety. As Reason (1990, p. 198) puts it 'the risk of an accident will be diminished if these pathogens are detected and neutralized proactively'. This ability to communicate risk is fostered by a shift from a paradigm of blame to one of inquiry with learning as the objective. It can become a pillar (one of many) on which foresight and proactive response is built into the maintenance of quality with respect to safety, security and environmental protection.

Based on the foregoing and guided by the overall inquiry into how and to what extent shipping companies learn from, filter and give credence/acceptability to differing risk perceptions and how this affects the work culture with special regard to group/team dynamics and individual motivation, this author has sought to explore in greater detail the outworking of these theoretical dynamics and how they affect organizational learning in the specific context of the maritime industry.

Context of Study

The need for quality/excellence – as manifested in the meeting of high standards in safety, security and environmental protection – is generally undisputed in the maritime industry. During the latter part of the twentieth century, there was a significant decrease in number of incidents, largely attributable to better and more comprehensive legislation relating to design, technology and human competence. However, it appears that this trend is now being reversed. According to Soma[6] (as cited by Richardsen, 2008):

> DNV's statistics shows that a ship is twice as likely to be involved in a serious grounding, collision or contact accident today compared to only five years ago. In addition, estimates show that also the costs of these accidents have doubled. Since this is the general trend for the international commercial [shipping] fleet, the maritime industry needs to act on this immediately.

This new trend (see Figure 1.2) is perceived to be mainly attributable to pressures on the manning situation from increased ship numbers, commercial goals and a shortage of ship officers (Hernqvist, 2008). The recent financial crisis seems to only have had a temporary effect on these pressures.

Consequently, despite significant advances in maritime technology, the role and importance of the human element[7] is again receiving increased attention. The maritime industry is about people. It is run by, serves and poses risks to people. Furthermore, similar to almost every other area of human endeavour, the industry seeks to group humans together in organizations and teams and to create structures that optimize the performance of these groupings. Obviously this necessary aggregation of the work force in social units has its potentially adverse effects.

A culture of excellence is significantly reliant on the behaviour of individual maritime organizations (be they educational institutions, shipping companies or administrations) in their day-to-day activities and the choices they make with regard to their human resource. One academic domain into which such matters fall is organizational theory with a special emphasis on organizational learning and behaviour. This area of study relates to how organizational culture is influenced by individuals, teams and the organization as a whole and what approaches/tools

---

6   Principal Safety Consultant in Det Norske Veritas (DNV) Maritime.

7   The term 'human element' is used advisedly in this context. As O'Neil puts it 'the phrase "the human element" is one of those terms that in a way helps to obscure the true meaning rather than to clarify it. When we speak of the human element or the human factor we are in fact talking about people and I think that we should always remember that our industry depends upon people, whether they are in the board room, on shore or on a ship on the high seas' (O'Neil, 2000).The term is used in this setting mainly because of its prevalence in IMO official documentation and with full awareness of the individuality that underlies such a term.

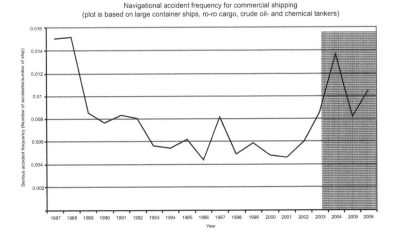

Figure 1.2   Accident frequency trends in maritime industry
Source: Det Norske Veritas (2009)

are needed to make an organization more effective in reaching desired goals. It helps to analyse consequences within a system, to attain an understanding of why individuals do what they do and to generate theories of action which can cope with change, manifestations of status, diversity and other multiple influences in and on organizations. The present study does not seek to explain the nature and functions of mental states but only to identify generic links between external variables in human behaviour in the socio-maritime context. It further seeks to explore and understand how maritime organizations try to avoid negative operational outcomes through learning and risk communication practices while recognizing that humans, organizations and their behaviours are extremely complex domains (International Maritime Organization [IMO], 1997).

The Shipping Industry: A Background

The shipping industry is recognized to be a complex one. Unlike many other industries – even in transport – it is characterized by significant diversity in a globalized setting. It is probably still the most international of industries, even in relation to the increased globalization[8] of other industries. The perils of the sea –

8   Giddens (1994, p. 4) sees globalization as being about the 'transformation of space and time' and defines it as 'action at distance'. In his opinion its intensification over recent years is due 'to the emergence of means of instantaneous global communication and mass transportation.' The definition of globalization proffered here is the extension of factors of production and the affected stakeholders and entities that impact the industry, from a

both regarding operations on the sea and management ashore – have challenged the ingenuity of man for centuries.

As at 31 December 2008, the world fleet of propelled seagoing merchant ships (tonnage greater than 100 GT) stood at 99,741 ships, equivalent to about 830.7 million gross tonnage and with an average age of 22 years (Lloyd's Register-Fairplay, 2009). International seaborne trade in 2007 was 7,420 million tonnes and 31,576 billion tonne-miles (Fearnresearch, 2009). Organisation for Economic Cooperation and Development (OECD) interests own about 68 per cent of the world fleet (Zachcial, 2008), but the majority of this tonnage is flagged in non-OECD countries. The largest registries (by tonnage) are Panama, Liberia and Bahamas (Lloyd's Register-Fairplay, 2009). It is generally accepted that the industry transports about 90 per cent of world trade in volume (International Chamber of Shipping [ICS], 2006; Warwick Institute for Employment Research, 2005).

The fleet is manned by around one and a quarter million people from different geographical regions – the rough equivalent of the population of Mauritius – with ships flying the flags of about 150 states (ICS2006; Warwick Institute for Employment Research, 2005). The majority of these are from Asia with the OECD contribution to manning decreasing significantly over the last two decades. The current situation is that most ships are manned by a multicultural crew drawn from almost every country of the world. These crewmembers, despite such extreme diversity and having to live in very limiting physical spaces for long periods, are expected to coexist in harmony and consistently work together to the highest levels of safety and efficiency.

The most significant regulations for the shipping industry at an international level emanate from the International Maritime Organization (IMO) and to a lesser degree, the International Labour Organization (ILO). Together these two bodies have issued a significant number of international conventions, codes, recommendations, rules and guidelines for regulating the industry with respect to maritime safety, pollution prevention, security, trade facilitation and labour issues. At the regional and national levels, regulations come from such bodies as the European Commission and sovereign states. Most of these nations regulate within the international framework as laid down by the IMO. It is noteworthy that despite the relatively high amount of legal instruments emanating from the IMO, the Organization, as such, has no enforcement powers. Such powers rest with the individual sovereign states, most of whom express this power by national legislation and enforcement by national maritime administrations.

In the shipping industry, an emphasis has been traditionally placed on approaches to safety/quality such as training, improved design of the man–machine environment or interface and control/restraint through legislation.

---

national/regional setting to a global one, i.e., an extension of the input parameters of an industry and the scale of its impact to a world wide level while still affecting local and personal social contexts.

Although these have been largely influential in the reducing level of accidents (and although legislation is recognized to be ubiquitous and indispensable) the place of other human factors (motivational, social, psychological and organizational) is increasingly acknowledged as critical to a further advancement of a paradigm of quality. The limitation of law[9] in achieving global goals of excellence in the maritime industry lies mainly in the reliance of law on negative consequences of deviation. Seldom does the law motivate based on a reward system (Schuck, 2000, p. 437). However, positive reinforcement has been found to be a high level motivator for subsequent behaviour. The law (as a form of social control and change) needs to be complemented by other mechanisms (Schuck, 2000; Vago, 2006, p. 347). Organizational factors that generate intrinsic motivation for excellence, as opposed to the extrinsic motivation of legal measures, are one such category of mechanisms and should arguably be given a lot more attention at organizational, national and international levels.

The organizing of the shipping industry is often a very international undertaking. Recruitment for the industry's human resource is mainly done at an international level and unlike for many other global organizations, shipping's organizational units are characterized by significant dispersion and mobility. A ship may be ordered by German interests; built in South Korea; trade between Australia and the European Union flying the Liberian flag and managed by an operational company set up in the United States. It may be then chartered to a Brazilian entity and carry cargo belonging to Chinese shippers for Danish buyers. The crew may be Filipino and Ghanaian seafarers employed by a manning agency located in Cyprus and have licences from the Philippines, Ghana or UK, and endorsed by Liberia. Bunkers may be acquired in Rotterdam and the crew changed in Bremen. Ship operation is also characterized by intense economic competition, decreasing manning levels and a relatively fast pace of technological change. This makes the industry even more vulnerable to risk as decision-makers often focus on 'short-term financial and survival criteria rather than long-term criteria concerning welfare, safety and environmental impact' (Rasmussen, 1997, p. 186).

Organizational control for quality purposes, therefore, is not a simple matter. Neither is the teamwork required to maintain operational excellence. This 'internationalization' of the shipping industry has been rapid and has occurred at the same time that the industry has maintained very conservative structures on board ship and ashore.

Traditionally, organizational structures on ships have been hierarchical. The standard structure for foreign-going ships is shown in Figure 1.3. While there are deviations (with some roles all but disappearing) the basic format used on most ships is very similar. This strict hierarchical structure has significant implications

---

9   By law (in this text) is meant public law which 'takes the form of an enforceable statute, an authoritative agency regulation or order, or a court decision' (Schuck, 2000, p. 419).

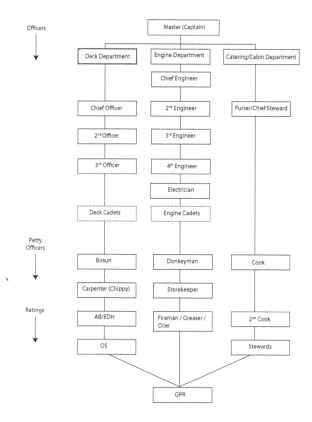

Figure 1.3   Standard traditional structure of shipboard organization

for risk communication which is sometimes exacerbated by the presence of different nationalities in the hierarchy.

There have been arguments for substantially different organizational structures on board ships (see Barnett, Stevenson and Lang 2005; Dyer-Smith, 1992). However, these have so far had limited impact on the industry as a whole. The status quo has its roots in centuries of shipping tradition. This high level conservatism is a marked feature of the shipping industry being evident in other contexts. The concepts of 'organizational learning' and 'learning organizations', for example, have gained extensive grounds in both academia and industry in other fields (Argyris and Schön, 1996, pp. xvii–xix). There has not been much discussion regarding this in the maritime sector. Generally speaking 'learning' in the maritime field has been legislation-driven. In other words, compliance with externally generated legal instruments seems to be the main motivation and engine for change in the industry as a whole.

The current work is premised on the notion that the achievement and maintenance of a culture of quality in terms of safety and environmental stewardship, requires shipping organizations to exhibit the characteristics of learning organizations. Contrary to the thinking which sees 'organizational learning' as an outcome – a change in strategy and action – this study sees it as more of a process which is cyclical in nature – an ongoing reflexive process (*reflexive* in the sense in which Giddens (1984, p. 3) defines it) which facilitates the achievement of necessary stability or necessary change.

The latter view makes it imperative that the analysis of individual and organizational knowledge be elevated above the level of task knowledge to consider social knowledge and how organizations rank various perceptions of risk. An important consideration is team learning and the factors that enhance or detract from this. This is in agreement with Edmondson (1999, p. 353) when she conceptualizes group learning as 'an ongoing process of reflection and action, characterized by asking questions, seeking feedback, experimenting, reflecting on results, and discussing errors or unexpected outcomes of actions.' The International Safety Management (ISM) code requires that companies 'establish safeguards against all identified risks' (IMO, 2002). This task is seen as increasingly important, especially in light of more demands for accountability and social responsibility in risk management. Granting that perceptions of risk differ and that there are differing epistemologies of risk in any set of stakeholders, how do organizations, in their quest to identify risks (and thence establish safeguards), give credence and acceptability to the different epistemologies of risk? A further question for examination is how this process of organizational credence-giving affects team psychological safety and ultimately organizational learning.

An organizational mindset that considers it necessary to integrate individual learning into group learning and subsequently into organizational learning, will value individual perceptions of risk and create the climate where individuals and teams would feel psychologically safe to communicate their perceptions of risk with the aim of arriving at a shared and considered group perception of risk. It is in this context that a similarity to the Social Amplification of Risk Framework (SARF) (Kasperson, Kasperson, Pidgeon and Slovic, 2003) at a micro level – team/organizational levels – is evident. The social dynamics that amplify or attenuate risk signals is just as pervasive in the micro-setting of a shipboard team (or a ship–shore team) as it is in the wider sense of the public and media.

Aims and Value

The focus of this book is how shipping companies as organizations learn from, filter and give credence/acceptability to differing risk perceptions and how this influences the work culture with special regard to group/team dynamics and individual motivation.

The next two chapters (Chapters 2 and 3) set out the theoretical underpinnings of the study and a review of the literature identifying trends in the fields of organizational theory, risk management and team/organization dynamics in the context of the maritime industry. From this synthesis of the literature a number of specific issues are introduced which are addressed using a mixed-methodology research approach. The methodology entailed the conducting of a survey of ship officers which sought global measures of team psychological safety, leader inclusiveness, organizational learning, worker engagement/perceptions of organizational support and other variables such as rank, age, gender and nationality. A second part of the methodology was undertaken via a qualitative study of six shipping companies sampled purposively based on the recent occurrence of serious accidents, organizational change and diversity. Details of the approach, questions and methodology are set out in Chapter 4. Chapter 5 presents the findings which are then discussed in Chapters 6 and 7. Chapter 8 concludes the study with implications of the findings and recommendations.

In seeking to address the global issue under consideration, this effort has had the following aims:

- Finding out how shipping organizations give hierarchical credence and acceptability to the risk perceptions of the various actors in the industry.
- Examining whether any such hierarchical acceptability of risk perception influences team psychological safety and worker engagement in the shipboard context.
- Reviewing the effects of the above in accident causation, recovery and prevention.
- Making a contribution in terms of optimizing the support context for teamwork – organizational behaviour, structure, resources and training.
- Helping to stimulate the consistent evolution of maritime organizations[10] into learning organizations.
- Discussing the effect of better team climate on attrition rates of seafarer numbers.
- Contributing to the enhancement of maritime education and training in the academic and industrial setting.

The book's focus is specific to the maritime industry and discusses the achievement of a learning culture in the context of quality operations. The term 'quality operations' refers to the consistent achievement of globally accepted goals of safety, security, pollution prevention and continual optimization of competence and motivation of people – operational excellence or exemplary performance, in other words. The necessary corollary of economics and profitability of operations (impact of quality on the 'bottom line') is acknowledged but it is beyond the scope

---

10   ... especially those shipping companies that do not contemplate learning and change outside the paradigm of external legislation. These are companies which may be said to operate in an 'evasive' or 'compliance culture' as opposed to a 'safety culture' (Mathiesen, 1996; Sudhakar, 2005) and which seem to form the bulk of contemporary shipping (Anderson, 2003).

of the current study to examine this in detail, except to note that achieving high levels of safe, secure and environmentally-friendly operations is a goal which is compatible with high economic performance (Mol, 2003).

Chapter 2
Social Dynamics and the Construal of Risk

The first step in the risk management process is to acknowledge the reality of risk. Denial is a common tactic that substitutes deliberate ignorance for thoughtful planning.

Charles Tremper[1]

Purpose and Outline

The focus of this study is how shipping companies as organizations learn from, filter and give credence/acceptability to differing risk perceptions and how this influences the work culture with special regard to group/team dynamics and individual motivation. This chapter is the first of two that set out the theoretical foundations for this work. It draws from quite diverse sources, an eclecticism deemed necessary given the multifaceted nature of the maritime challenge being addressed and the view that risk and management issues must be treated in a cross-disciplinary manner (Mullins, 2005, p. 19; Rasmussen, 1997, p. 183; Renn, 1992, p. 54). The theoretical frameworks underlining the studies are adaptations of the Social Amplification of Risk Framework (SARF) (R.E. Kasperson et al., 1988), Organizational Support Theory (Eisenberger, Huntington, Hutchison and Sowa, 1986) and Social Exchange Theory (Homans, 1958). These frameworks and the resulting review are drawn mainly from the literatures in the theoretical disciplines of social-psychology,[2] management and organizational theory, risk management and accident causation/modelling. Also addressed in this chapter is a discussion of the structure versus agency debate regarding organizational research. In all of these domains, the focus is on only those issues relevant to the current study. Further specific links to theory are critically discussed in later chapters discussing the research findings and subsequent implications, conclusions and recommendations.

---

1   Retrieved 30 May 2009 from http://thinkexist.com/quotation/the_first_step_in_ the_risk_management_process_is/202055.html.

2   One of the meanings offered by Murray et al. for 'psychology' in the *Compact Oxford English Dictionary* (1991, p. 1461) is 'the study of the behaviour of an individual or a selected group of individuals when interacting with the environment or in a given social context'. This is what is intended in this text, with an acknowledgement of the cognitive aspects of such behaviour.

Social psychology is the scientific study of how people's thoughts, feelings, and behaviours are influenced by the actual, imagined, or implied presence of others (Allport, 1985, p. 5).

Risk

During the last century, there has been an increasing focus on risk and its management (Beck, 2009; Luhmann, 2006; Slovic and Weber, 2002). This is partly due to the fact that many societies are decreasing their reliance on mysticism and religion while increasing reliance on technology and human decision-making systems to deal with the future (Bernstein, 1998). Fundamental to this trend, is societies' uncertainty about the future and the desire to have it aligned with preferred expectations (Luhmann, 2006; Mythen, 2004) together with the perceived increased potential for recreancy[3] on the part of individuals and organizations, the decision-making systems (Freudenburg, 1993; Pidgeon and O'Leary, 2000; Vaughan, 1996, 1999a, 1999b). With the increase of focus on risk has come significant debate about the concept of risk (Campbell, 2006; Fischhoff, Watson and Hope, 1984) – what it is and how it can be assessed and managed.

The maritime industry is also concerned with risk and rightly so, given that the word itself has etymological origins in the nautical world (Skjong, 2005, p. 40) or at least was first applied consistently and significantly in maritime navigation and trade (Luhmann, 2006, pp. 8–10; Oppenheim, 1954, p. 9). The Italian derivation, *risicare*, has connotations meaning 'to dare' (Bernstein, 1998). The question that contemporary scholarship is finding more and more relevant is who dares on whose behalf and on what basis?

The International Maritime Organization (2002a, p. 4) defines risk as 'the combination of the frequency and the severity of the consequence'. This is similar to the definitions given by a significant number of organizations/publications from risk-sensitive industries including, the European Environmental Agency (EEA), Food and Agriculture Organization Emergency Prevention System (FAO-EMPRES), the International Atomic Energy Agency (IAEA), the International Organization for Standardization (ISO)/International Electro-technical Commission (IEC) Guide 73 on Risk Management Vocabulary, and the Society for Risk Analysis (SRA) among others.[4]

Definitions such as the above are premised on the belief that risk is objective, determinable and quantifiable, leading to the notion that, though perceptions of this objective risk may differ, the risk – essentially rational and objective – remains the

---

3   'This usage draws on one of the two dictionary meanings of the term, namely a retrogression or failure to follow through on a duty or trust. It is intended to provide an affectively neutral reference to behaviors of persons and/or of institutions that hold positions of trust, agency, responsibility, or fiduciary or other forms of broadly expected obligations to the collectivity, but that behave in a manner that fails to fulfill the obligations or merit the trust' (Freudenburg, 1993, pp. 916–17).

4   For the full texts of these definitions and other more subjective ones, see Annexes B and C of the White Paper on Risk Governance by the International Risk Governance Council (Renn, 2005, pp. 139–54) and the ISO Guide 73 (International Organization of Standards (ISO) and International Electrotechnical Commission (IEC), 2002).

same. This positivistic ideal is countered strongly by some who believe that risk in essence is subjective and that all notions of risk are therefore relative (Otway, 1992; Renn, 1992; Sjöberg, 2006; Slovic, 1992, 1999). Slovic, for example, states that:

> Much social science analysis rejects this notion [of objective quantification of risk assessment], arguing instead that risk is inherently subjective. In this view, risk does not exist 'out there,' independent of our minds and cultures, waiting to be measured. Instead, human beings have invented the concept risk to help them understand and cope with the dangers and uncertainties of life. Although these dangers are real, there is no such thing as 'real risk' or 'objective risk.' The nuclear engineer's probabilistic risk estimate for a nuclear accident or the toxicologist's quantitative estimate of a chemical's carcinogenic risk are both based on theoretical models, whose structure is subjective and assumption-laden, and whose inputs are dependent on judgement. (Slovic, 1999, p. 690)

For those in the latter camp, an objective measuring of risk is impossible. In what may be considered an extreme view, Sjöberg (2006, p. 692) even describes such a measuring of risk as 'a chimera'. The arguments are not unlike the discussion regarding the nature of social science research and human reliability analysis – the decades old debate about the suitability of quantitative versus qualitative/heuristic methods when it comes to the scientific analysis of 'human factors' (Becker, 1996). As Bernstein (1998, p. 6) notes, there is a:

> persistent tension between those who assert that the best decisions are based on quantification and numbers, determined by the patterns of the past, and those who base their decisions on more subjective degrees of belief about the uncertain future. This is a controversy that has never been resolved.

Rosa's (2003) somewhat middle ground position with respect to risk, combines objectivity with the uncertainty of future events inherent in humanity. He defines risk as 'a situation or an event where something of human value (including humans themselves) is at stake and where the outcome is uncertain' (p. 56). This definition subsumes the notion of 'value', 'potential outcome' and the 'uncertainty' of that outcome that is found in almost all other definitions. It affirms an ontological state for risk but allows for a hierarchical arrangement of the epistemology of risk with the latter's consequential dependence on subjectivity. The definition also allows for the increased robustness of the concept of risk, to include instances where the outcome may be positive such as 'investment risk' and 'thrill risk' (Rosa, 2003, p. 60). Waring and Glendon refer to 'investment risk' as 'speculative risk' and differentiate it from 'pure risk', where the potential outcome is unwanted and must be eliminated, reduced or controlled (Waring and Glendon, 1998, p. 4). This distinction is important, because often top-level decision-makers in companies may have a view of 'pure risk' that is tempered by a business world in which 'speculative risk' dominates; the 'rational actor' as opposed to the 'psychometric' paradigms of risk. Similarly 'thrill risk' has been noted to have been a causal factor in accident causation.[5]

For Hansson (2004, p. 10) there are a number of specialized uses and meanings of risk in technical contexts.[6] He indicates that five of these are particularly important since they are widely used across disciplines:

1. Risk = an unwanted event which may or may not occur.
2. Risk = the cause of an unwanted event which may or may not occur.
3. Risk = the probability of an unwanted event which may or may not occur.
4. Risk = the statistical expectation value of an unwanted event which may or may not occur. [The expectation value of a possible negative event being the product of its probability and some measure of its severity].
5. Risk = the fact that a decision is made under conditions of known probabilities ('decision under risk') [as opposed to 'decisions under uncertainty'].

Möller, Hansson and Peterson (2006, p. 421) are of the opinion that definition 4:

> may be considered as a compound of (1) and (3), where the unwanted event has been assigned a value (disutility) and a probability. However, there are aspects of (1) and (3) that are not very easily included in the representation of (4). Notably, (4) rests on the presupposition that an unwanted event can be given a precise value, which is a strong assumption not needed in (1) and (3) ... We shall assume that probability and harm [consequence] are the major components of risk, but we do not need to assume that they can be combined into a one-dimensional measure of risk as in definition. (4)

Both one and two of Hansson's definitions have a qualitative sense whereas three and four connote some quantitative measure. The latter couple seem to be more appropriate in risk-based decision-making where they give a comparison between options. They do not appear to be convincing definitions in absolute terms.

Some texts (for example, Kuo, 2007, pp. 47–52) differentiate between risk and danger/hazard.[7] The latter is defined more like Hansson's definitions one and

---

5   See for example the crash of Pinnacle Airlines Flight 3701 Bombardier CL-600–2B19, N8396A Jefferson City, Missouri on 14 October 2004. Full report at http://www.ntsb.gov/publictn/2007/AAR0701.pdf retrieved 1 December 2008.

6   Hansson acknowledges the existence of other technical definitions which are well established in specialized fields of inquiry.

7   Other texts even differentiate between danger and hazard with the Society of Risk Analysis (SRA) defining 'danger' as expressing 'a relative exposure to a hazard. A hazard may be present, but there may be little danger because of the precautions taken' (Renn, 2005).

two and definitions three and four more rightly as risk. A tanker spilling oil may be seen as an objective hazard. The cause of this spill may be a collision and the collision itself may well be considered an objective hazard. The risk element of this is the probability (whether determined a priori or based on statistical analysis of homogeneous data) that this will occur and the potential severity of such an occurrence at various levels if indeed it occurs. In theoretical terms, such a risk can be objectively determined in terms of a statistical expectation value or calculated in some other way if values are given for 'probability' and 'consequence'. While this may satisfy the mechanics of positivistic science – and 'absolute rationality' (Perrow, 1999, pp. 315–16) – in its treatment of pure risk and may be useful for a pragmatic assessment phase of risk management, it is not that helpful when it comes to the acceptance phase (decision-making and policy-setting). This phase of risk management has to assess relative risks, components of speculative risk and options for risk mitigation as well as 'ex-post responsibility for risk' (Thompson, 1992) and the binary nature of risk outcomes in individual/societal lay perceptions (Botterill and Mazur, 2004; Jackson and Carter, 1992). Jackson and Carter further argue that 'the minimization of risk through increasingly rational behaviour is an unattainable goal and that since perception determines what is rational, we need to concentrate on perception rather than rationality' (p. 41). Furthermore other social issues, such as ethics, exposure, immediacy, morality, consciousness, intent (running a risk or taking a risk) and various other subjective values become important. While scientific methodology may exclude data that is not statistically significant[8] (and by the very nature of scientific inquiry this may be required), in decision-making such data may be highly relevant. Statistical outliers are often high impact, though rare and determine the genesis and sustenance of influential policies (Taleb, 2007).[9]

To elaborate further with the tanker example, a determination of the risk of a tanker spilling oil in a particular location may be calculated from the aggregation of risk contributing factors such as proximity to shore, data from previous accidents and near-misses relating to tanker groundings, weather conditions, age and kind of tanker, nature of sea passage, traffic density and so on, together with quantification models of the consequences.[10] As discussed earlier, this will give what is regarded by many as a pragmatic and objective indication of risk. The determination of this risk will then mean that thresholds of tolerable risk can be set. An opposing subjective view will then argue that, there is a quantum of other factors, both acknowledged/stated and otherwise, which are real and pertinent

---

8   The statistical significance of relationships in a data set is associated with the judgements given about the degree to which the results are attributable to chance.

9   No one disputes the 'rareness' of an event (based on historicity) like the 9/11 WTC attacks in the US (2001) nor the significance or impact – economic, legal or social – of that one event.

10   See for example the use of such a formula from a suggested risk determination process in the EU project MarNIS (Glansdorp, 2009).

in an existential sense, but which are ignored in the quantitative models. These factors cannot be fully assessed by even experts but have to be considered – no matter how rudimentary that consideration is – in any 'determination' of risk. A key contentious issue relates to the derivation of probability figures based only on limited sets of historical data. For example, a risk parameter based on the place of a vessel on a list derived from previous administrative encounters (for example, Port State Control white, grey and black lists) is seen as not adequately addressing the risk of a major accident which happens for the first time, which risk is real but may not be captured by the existing epistemic level[11] or approach to data collection; in effect the epistemic level that has been given credence by dominant societal constituencies.[12] Another key consideration would be what constitutes 'value' and 'consequence' both in identity (what) and in severity. What are the full consequences of a tanker spilling oil in a confined region? Is it possible to include all losses without limitation in proximity or time? What constitutes 'consequences' is significantly reliant on individual perceptions and perspectives. In this sense the risk we can humanly 'objectively' determine, will always be 'subjective' because it is restricted by our particular epistemic limitations, that is, the limitations in our knowledge, not only of what factors go into the specific risk but also how much we know of these factors. It therefore follows that the so-called objective determination of risk becomes tied in to the subjective preferences and values of socio-political constituencies.

By extension, there will also be subjective preferences/differences in what constitutes adequate risk mitigation measures. The rationality associated with the managerial assessment of risk in high-risk industries may only be 'bounded rationality'[13] or 'limited rationality' to the extent that it is dependent on specific contexts. Accordingly 'fixing the disaster threshold is almost at the discretion of the person arguing in these terms' and contexts (Luhmann, 2006, p. xxx). Like risk determination and acceptance, risk mitigation approaches will also be subjective. Going back to our tanker example, even if different rationally-bounded constituencies agree on what the risk is[14] (in terms of probability and consequence), there will arise differences in how to mitigate that risk. Some constituencies believe that tanker safety, can arguably be better improved with

---

11   Or claims to knowledge (see Giddens, 1990, p. 54).

12   There are many other limitations of such restrictions in data and in recognition of this, some scholars and delegates to the IMO have argued that Port State Control data, for example, may show no significant correlation to accident data and that there is the need to harmonize the two sets of data for a more comprehensive picture of risk (Degré, 2008; IMO2008).

13   The term 'bounded rationality' was initially used by Herbert Simon in another context (Simon, 1982) but is used here in line with Perrow's statements on 'absolute, bounded and social/cultural rationality' (Perrow, 1999, pp. 315–23).

14   Note that such an agreement is not necessarily a determination of the 'real' risk (which may or may not objectively and ontologically exist) but the merger, across different levels of human aggregations, of different perceptions of risk.

other risk mitigation technologies and policies, but based on what many described as a political and hasty process, the double hull option[15] is the only one accepted by most jurisdictions (Mukkadayil, 2001).

The process of the construal of risk then, as well as how risk can be mitigated, becomes a manifestation of the degree to which risk communication and consultation are effective in a particular organization, society or on a global scale. It also indicates the credence that is given to different epistemologies of risk by that society. While it may be impossible to avoid the power/class dynamics associated with these processes and the accompanying imbalances created by such 'power games', the necessity of seamless risk communication (or rather 'risk perception communication') is obvious. It is needed across all phases of the risk management process. The challenge is in how to optimize these processes for pragmatically defined benefits. In the words of Luhmann (2006, p. xxix):

> How is social consensus (or even a mere temporary common basis for communication) to be achieved if this is to take place within the horizon of a future about which – as everyone knows – one's interlocutor, too, can express himself only in terms of probability and improbability?

These social dynamics that are implicit in conceptualizing risk make 'agency' an issue of importance. If our notion of risk, founded on epistemic limitations, are necessarily subjective, then all risk we perceive will emanate from a decision or a lack of a decision, based on whether we know enough, care enough, or act enough.

These arguments/discussions, though primarily related to macro-societies, are just as applicable to smaller groups and the individuals within them.

There are merits/demerits in both sides of the argument – subjective or objective (Renn, 1992) – but the necessity of attempts at risk communication cannot be denied by either. In Luhmann's view (2006, pp. xxxi–xxxii) it is not the place of the social sciences to take sides, let alone decide the issue, but to examine the effect of these views on society itself – 'find out more exactly what is going on' as a normal functioning of society. The view taken here – and considered appropriate in the specific maritime operational research context – is what will be termed an 'unbounded usage' of 'risk'.[16] Risk, in essence, is a social construct and in that sense completely subjective. However, any particular and specific risk can be claimed to be objective reality ontologically (Rosa, 2003) and not necessarily dependent on the consciousness of decision-makers and those affected by decisions (Giddens, 1990, pp. 34–5) especially at the aggregated level of macro-society and

---

15　　Other options were e.g. the Mid-deck and Coulombi Egg ship designs (see IMO MARPOL Annex 1 Regulations 13F(5) in the 2002 consolidated edition of MARPOL and Annex 1, Regulation 19 (5) in the 2006 consolidated edition and MEPC Circ. 336) together with other non-design risk mitigation schemes. See some views on http://heiwaco.tripod. com/professionnels.htm retrieved 18 June 2009.

16　　Hansson (2007, par. 2) prefers 'non-regimented usage' of risk.

risk governance. It may also be seen as being scientifically objective in the sense that it is a specific 'expectation value' of an uncertain event or the cause of such event based on current and accessible knowledge and in a context of bounded rationality. However, in an unbounded context, the lack of knowledge about the existence – or full dimensions – of any outcome and all of its potential causative factors and consequences, makes the associated risk indeterminable objectively. The 'uncertainty' element of human knowledge regarding this risk makes it essentially subjective, that is, risk is epistemologically subjective (Rosa, 2003). Additionally, for risk to be 'present' there must be exposure – the consequences of a proposition that is (or would be) interpreted by the individual/entity (often with affective connotations) as needful of care – and this again is subjective. Even where risk is quantitatively specified as a probability, it has to be borne in mind that the very essence of probability is the notion of uncertainty (whether epistemic or aleatory) which is almost by definition subjective/perspectival. Furthermore the reliance of frequentist probability on 'historicity' creates a problem for predictiveness. In estimating risk, a healthy amount of foresight and abstraction is required. As Sagan[17] (1993, p. 12) notes:

> Things that have never happened before happen all the time ... There must be a first time for every type of historical event that has occurred in the past, and the lack of accidents is therefore insufficient evidence for making strong statements about future possibilities.

In summary this work takes the critical realist[18] view that particular risks are objective ontologically. However, the epistemology of risk is subjective. Risk perception (of individuals, organizations and societies) will lie on a continuum with two extreme ends marked by different levels of uncertainty. One end is purely aleatory (minimum or no knowledge) and the other end is fully shared and congruent societal epistemology of a particular risk.[19] It is noteworthy that the latter is not the same as 'full knowledge'. In between these two extremes will lie the particular 'subjectivity' of risk, that is, the perception of risk based on a

---

17   Sagan's text is in reference to the possibility of nuclear accidents. Tools of prediction based on historicity or the status quo – such as the 'bell curve' – are severely limited when it comes to inference to predict rare but high impact events (the statistical outliers). See Nassim Nicholas Taleb's arguments in 'The Black Swan: The impact of the highly improbable' (Taleb, 2007).

18   Critical realism, following Trochim and Donnelly, is defined here as 'the belief that there is an external reality independent of a person's thinking (realism) but that we can never know that reality with perfect accuracy (critical)' (Trochim and Donnelly, 2007, p. 19).

19   See Bernstein for a similar analysis with reference to pure games of chance and games whose outcomes depend to a degree on skill (a qualitative value) and for an interesting discussion of the origins of probability (Bernstein, 1998, pp. 14, 48–9).

number of qualitative dimensions that make risk a multidimensional construct (Fischhoff et al., 1984; Hansson, 1989; Jenkin, 2006).

A list of such qualitative dimensions of risk is given in Table 2.1. The dimensions in italicized text appear to vary directly with risk perception while the others appear to vary inversely with risk perception.

Table 2.1   Qualitative dimensions of risk

Voluntariness	**The extent to which exposure to the hazard is voluntary**
Immediacy	The time frame within which consequences are first noticed or felt
Temporal scope (duration)	The scale of consequences in terms of time, that is, the time frame within which consequences stay (for example, a year, a decade, a generation)
Spatial scope	The scale of consequences in terms of space, that is, the geographical area covered by effects of exposure (for example, national, global and so on)
Knowledge and comprehension	The individual or entity's degree of awareness and understanding of exposure based on education, information flow and so on
Vulnerability	The perception of degree of vulnerability exposure brings
Expert knowledge	The degree to which experts know the hazard
Controllability	The extent to which a victim can control the severity of consequences due to exposure. This dimension subsumes perceptions of 'preventability' as well as the 'mitigation of consequences' after the event
Novelty (familiarity)	The extent to which the hazard is new/familiar to the individual or society
Catastrophic potential	The number of fatalities/damage as it relates to time
Dread	The extent to which the effects of exposure are dreaded
Increasing/Decreasing	The change in severity of consequences over time
Equitability (fairness)	The sense of the balance between costs and benefits of the risk or hazard (for example, in terms of physical, economic, health, social costs versus benefits)
Personal impact	The extent to which the risk affects the entity personally
Observability (ease of detection)	The extent to which the effects of exposure are observable/ perceptible or easily detected
Source, nature and type of hazard	Origins of the risk for example, whether natural or artificial; characteristics of hazard and whether there is a history or stigma

These – and other dimensions (Robertson and Stewart, 2004) – are in turn influenced by societal and individual beliefs, values, ethics, morality, trust and so on.

In developing a predictive risk management system, what is at stake is how to validate the resulting hierarchical epistemology of risk and to communicate this validation to the satisfaction of all stakeholders. This view is considered relevant to organizational pragmatism.[20]

Others hold different views from the position taken in this study (especially based on whether their view of probability is extremely Bayesian or Frequentist) and arguments persist as to whether risk (and even probability) is objective or subjective (Campbell, 2006; Holton, 2004; Thompson, 1990, 1992; Valverde, 1991). Considering that it is virtually impossible in unique complex systems to enumerate and fully describe/quantify the elements of an unwanted event (or its causes/consequences, exposure and uncertainty) in absolute technical and social terms, there will always be a subjectivity to adverse risk based on knowledge and so on. which will create 'epistemic uncertainty' (Möller et al., 2006, p. 421) – and a disparity of perceptions between individuals, groups and other stakeholders at different levels. Any process of risk governance should consider the more measurable 'factual' elements of risk as well as the more 'subjective' and 'socio-cultural' elements (Renn, 2005, p. 12). The importance of the latter is almost universally acknowledged.

Risk Perception and Communication

The *New Oxford Dictionary* (Pearsall, 2001) gives a psychological definition of 'perception' as 'the neuro-physiological processes, including memory, by which an organism becomes aware of and interprets external stimuli'. It has also been described as the 'psychological process by which we select, arrange and understand information' and which 'defines reality for the perceiver' (E. Wilson and Rees, 2007, p. 123) and as non-voluntary responses to stimuli for example, visual and auditory stimuli (Meeker, 1971, p. 485). A comprehensive definition is given by the *Merriam-Webster Dictionary*: 'a mental image; the integration of sensory impression of events in the external world by a conscious organism especially as a function of non-conscious expectations derived from past experience and serving as a basis for or as verified by further meaningful motivated action' (Gove, 1993). In this context, 'risk perception' is considered to subsume, not just sensory elements that may give an indication of risk[21] but also attitudes and expectations

---

20   Pragmatism here is not a reference to the philosophical worldview but is used to refer to a practical point of view or practical considerations. http://dictionary.reference. com/browse/pragmatic retrieved 17 July 2009.

21   Risk cannot be sensed, being in essence bound to the future. Sensory mechanisms (the five senses) are only associated with the present and so risk perception (defined as

regarding the construed object – in this case risk (Hansson, 2007; Sjöberg, 2000, p. 408). It has been defined as 'the risk we envisage, which results from how we assess the chance of a particular type of accident happening to us and how concerned we are with such an accident' (Risk Research Committee, 1980; Marek et al., 1985, 1987 as cited in Rundmo, 1992).

This goes beyond 'risk cognition' and involves attitudes and beliefs – affective states that are subjective in nature but impact human behaviour and responses in the face of danger. Not only do such perceptions exist, but they are held to be real by those who hold them, no matter how irrational, stereotypical or distorted they may seem to others (Hale and Glendon, 1987; E. Wilson and Rees, 2007). As Bandura notes 'people's ... affective states, and actions are based more on what they believe than on what is objectively true' (1997, p. 2).

When individuals become part of a group, these perceptions may persist, but they are significantly influenced by the social (for example, organizational) environment. According to E. Wilson and Rees:

> The informal modus operandi of an organization, which includes factors such as history, policies, communication networks, and humour, can introduce major biases into the process of social perception ... The process of perception is influenced, for better and for worse, by factors that are embedded in the existence of the organization and not just the attitudes and beliefs of employees. This perspective again draws us towards the representation of perception as a process in which a wide range of individual and organizational factors interact. (2007, p. 143)

It is clear that where team formation is envisaged, organizational dynamics and support systems should be such as to facilitate the consideration of varying perceptions of risk with regard to quality-goal-attainment. Communication and sharing of perceptions of risk therefore become important in a team setting. This may not be critical for groups (as differentiated from teams) since they are not – by definition – necessarily formed to pursue specific goals.

The dynamics of a team or societal setting may cause varied risk perceptions to converge, for example where the dominant perception attributable to say social standing or expertise becomes the accepted societal notion of risk over time. The convergence of risk perception could hypothetically be high or low at the macro-level of society.[22] Either convergence (high or low) has its advantages and disadvantages. At the micro-level of teams the existence of diversity in risk perception can be an asset since the process of attempted convergence may be optimally used for teamwork. This process of convergence is critical to learning

---

sensory awareness) can only refer to perceptions of external stimuli that are linked to risk, based on memory, experience, education etc. and anticipation of the future.

22   The achievement of consistent societal convergence (macro-level) of high and low risk perceptions is considered neither possible nor desirable.

and is differentiated from the kind of consensus that characterizes 'groupthink'. The 'dialogue' of risk perceptions, and the attempted convergence that follows it, is important as 'no party [no matter how experienced or competent] has access to final truth with regard to risk and hazards; diversity is an asset' (Sjöberg, 2006, p. 683).

When this is considered against the background of recent research indicating that 'there is significant effect upon perceptions of risk of nationality' and that 'a worker's background and place within an organization do have an impact upon the perception of risk' (Bailey, Ellis and Sampson, 2006, p. 53), it becomes clear that the multicultural diversity apparent in shipping – which is often presented as a unique challenge and problem source – may really be an advantage to the maritime industry.

The Social Amplification/Attenuation of Risk

The effects of risk are not limited to the risk signal or risk event itself, but can have considerable secondary impacts based on dynamic social processes of amplification or attenuation. The literature in the domain traditionally emphasizes 'amplification' as opposed to 'attenuation'. The latter is, however, equally important. To avoid the tendency of referring to 'amplification' without mentioning 'attenuation' this study uses the neologism, 'amplenuation'.

The Social Amplification of Risk Framework (SARF), as shown in Figure 2.1, was introduced by Kasperson et al. (2003) and describes how 'risk and risk events interact with psychological, social, institutional, and cultural processes in ways that amplify or attenuate risk perceptions and concerns, and thereby shape risk behaviour, influence institutional processes, and affect risk consequences' (p. 2). Such secondary impacts could include stigmatization (of roles, organizations, modes of operation), regulatory impacts, financial loss and in the case of the maritime industry, also criminalization and high attrition rates of seafarers. SARF is a framework describing a social dynamic and not a theory as such and adequately accommodates a number of risk perspectives including the following approaches: actuarial, toxicological/epidemiological, probabilistic risk analysis, rational actor/ economic, psychometric, social theories and cultural theories (Renn, 1992; Slovic, 1992). Table 2.2 shows some of these different approaches to risk with merits and demerits.

Rosa's (2003, p. 63) use of the concepts of Hierarchical Epistemology and Realist Ontology (HERO) in the Social Amplification of Risk Framework (SARF) is significant in establishing a theoretical basis for research into organizational behaviour with respect to risk analysis. Lay ontologies are 'real' to operators in high-risk industries and the sharing and communication of risk is definitely impacted on by how organizations, through their systems, procedures and cultures, give credence and acceptability to the perceptions of risk that the different actors

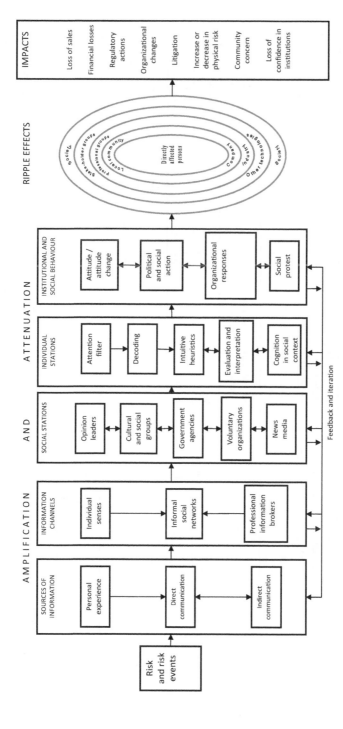

Figure 2.1   The Social Amplification of Risk Framework (SARF)
Source: Kasperson, J.X., Kasperson, R.E., Pidgeon, N. and Slovic, P. (2003)

Table 2.2   A systematic classification of risk perspectives/approaches

Source: Renn, O. (1992). Concepts of risk: A classification

	Approaches that integrate different perspectives (for example, social amplification of risk)						
	Actuarial approach	**Toxicology/ epidemiology**	**Probabilistic risk analysis**	**Economics of risk**	**Psychology of risk**	**Social theories of risk**	**Cultural theory of risk**
Base unit	Expected value (EV)	Modelled value	Synthesized expected value	Expected utility (EU)	Subjectively expected utility	Perceived fairness and competence	Shared values
Predominant method	Extrapolation	Experiments / Health surveys	Event and fault tree analysis	Risk benefit analysis	Psychometrics	Surveys / Structured analysis	Grid-group analysis
Scope of risk concept	Universal / One-dimensional	Health and environment / One-dimensional	Safety / One-dimensional	Universal / One-dimensional	Individual perceptions / Multi-dimensional	Social interests / Multi-dimensional	Cultural clusters / Multi-dimensional
	Averaging over space, time, context			Preference aggregation		Social relativism	
Basic problem areas	Predictive power	Transfer to humans / Intervening variables	Common mode failure	Common denominator	Social relevance	Complexity	Empirical validity
					Policy making and regulations / Conflict resolution (mediation) / Risk communication		
Major application	Insurance	Health / Environmental protection	Safety engineering	Decision-making			
Instrumental function	Risk-sharing	Early warning / Standard setting	Improving systems	Resource allocation	Individual assessment	Equality fairness / Political acceptance	Cultural identity
Social function	Assessment	Risk reduction and policy selection (coping with uncertainty)		Political legitimation			

have. By extension, team psychological safety, worker engagement, organizational learning and quality goal achievement are affected.

In the social amplenuation of risk, stakeholders can be considered to be 'risk alerters', 'risk mockers/deniers', and many others characterized by apathy. These categories may exist despite (or because of) the competence, experience or social background of individual stakeholders and be implicitly present even when not overtly stated. At group and organizational levels, good teamwork will seek to create a forum where there is no apathy and there is a convergence of views based on considered professional, competent and factual analysis. Subordinates in teams – especially in high-risk teams where competence is stressed – are perceived as often leaving risk determination and mitigating action to superiors and as exhibiting 'destructive obedience' (Sampson, 2002, p. 4). A different mindset will lead to more situational awareness of the working environment.

In standard ship teams most of these factors are brought to bear in a potentially high-risk environment. Optimum teamwork would suggest that team members (onboard and ashore) will be disposed to keep communication lines open no matter the hierarchical structures in place with the view of having a team-based perception of risk that optimally considers all potential dangers. Taking account of the social, organizational and individual factors that may inhibit or promote this communication is important in the intensely diversified world of shipping. In other words, if the ship-related team (onboard and ashore) is visualized as a microcosm of the Social Amplification of Risk Framework, then understanding the variables and dynamics at work will help determine best efforts at individual, team and organizational levels for optimizing teamwork, in the same way that understanding the macro-level implications and dynamics of the social amplenuation of risk should help to create a forum/reference for the analysis of policies for risk governance – risk appraisal, avoidance, control and consequence mitigation.

The dynamics of the social amplenuation of risk are fundamentally a communication issue and the filters that inhibit or permit/promote it. In the context of the operation of ships, one such filter is the construct of 'team psychological safety' which is discussed in the following section.

Team Psychological Safety

Cognitive workload in ship operation has been shown to be high and becoming increasingly so (Dyer-Smith, 1992; Lee and Sanquist, 2000). On the bridge of ships, for example, better technology implies that though there are less direct operational actions, there are more supervisory and decision tasks and also a requirement to divide attention between primary tasks such as navigation and secondary tasks such as engine and cargo monitoring. Hockey, Healey, Crawshaw, Wastell and Sauer (2003, p. 262) indicate that higher levels of threat in one task lead to impaired performance in secondary tasks.

Given these factors, the need for teamwork, long acknowledged in the maritime industry, is becoming more acute in contemporary times. The nature of teamwork is also changing. It was previously the case that teamwork meant that individuals focused on their specific tasks. However, the evolution of shipboard work seems to require more multitasking and decision support elements in teams.

Conditions that require increased communication across ranks, departments and even location (as in ship/shore relations) are being exacerbated by the prevalence of multinational crews, lower manning levels as well as the strict and continued adherence to traditional hierarchical structures among others. These are all factors which have the potential to inhibit critical communication in this high-risk industry.

Teams are 'work groups that exist within the context of a larger organization, have clearly defined membership, and share responsibility [to varying degrees] for a team product or service' (Edmondson, 2003, p. 257). They are made up of 'people who need one another to produce an outcome' (Senge, 2006, p. xiii).

Kozlowski and Ilgen (2006, p. 79) give a set of useful descriptors in defining a team:

- Two or more individuals [therefore includes dyads].
- Socially interact [either face-to-face or remotely for example, virtually].
- Possess one or more common goals.
- Brought together to perform organizationally relevant tasks.
- Exhibit interdependencies with respect to workflow, goals and outcomes.
- Have different roles and responsibilities [may be with respect to timing, function or location].
- Are together embedded in an encompassing organizational system with boundaries and linkages to the broader system context and task environment.

The effectiveness of teams is an area subject to significant research (see for example Amason, Hochwarter, Thompson, and Harrison, 1995; Barnett et al., 2003; Gibson, Zellmer-Bruhn and Schwab, 2003; Kozlowski and Ilgen, 2006; McGrath and Tschan, 2004). In a review of 50 years of research into the behavioural aspects of teamwork, Kozlowski and Ilgen (p. 78) point out that over the last five decades there has been a shift in the research focus from the study of 'small interpersonal groups' to the study of 'work teams' who are not necessarily co-located – shifting from 'individual jobs in functionalized structures to teams embedded in more complex workflow systems'. Arguably this shift is not pronounced in the maritime literature. The industry retains a team context that is still focused on small interpersonal team interaction on board ship. However, there are signs that more complex forms of teamwork that is not dependent on co-location are gaining research attention in shipping. The ISM requirement of a direct link between ship and highest level of shore management – the Designated Person Ashore (DPA) – is an example of this trend, a trend increasingly made possible by improved technological possibilities in telecommunications, that is to say, better technologies at lower costs. Both approaches to teamwork research, therefore, are pertinent to the maritime industry.

The literatures also have different emphases on the different variables perceived to impact optimum team relations. While some studies point to the importance of such variables as structure, resources, rewards and the environment as key to effective teamwork, others have emphasized cognitive and interpersonal factors and individuals' perceptions of these factors as critical (Druskat and Pescosolido, 2002; Hofmann, Morgeson and Gerras, 2003). The research premise in this study is that both of these approaches are important and each should not be considered without reference to the other. Cognitive and interpersonal factors are often evidence of – or based on – more tangible and measurable parameters such as structure and available resources.

The construct of psychological safety has been described in relevant studies as 'feeling able to show and employ one's self without fear of negative consequences to self-image, status, or career' (Kahn, 1990, p. 708). In the context of teams, where certain social influences, behaviours and perceptions can detract from optimum team performance, Edmondson (1999, p. 350) defines team psychological safety as 'the shared belief held by members of a team that the team is safe for interpersonal risk-taking'. This connotes that members have mutual trust and respect and are confident that 'the team will not reject, embarrass, or punish team members for speaking up' (Tynan, 2005, p. 229). Team psychological safety – or dyadic psychological safety – is pertinent because it has been shown in other industries to be correlated to worker engagement, the communication and sharing of perceptions of risk and ultimately with organizational learning (Edmondson, 1999; Tynan, 2005). This correlation is arguably mediated by organizational credence-giving and acceptability of risk perception. Team psychological safety is critical to the achievement of an organization's quality goals to the extent that the achievement of those goals is dependent on organizational learning. The term has been equated to 'group cohesiveness' but this is an equation that is flawed. The two are not the same. Team psychological safety may lead to group cohesiveness, but it is not the only possible determinant of group cohesiveness. The latter may also be the result of unquestioned consensus and may manifest in 'groupthink'. The term groupthink is attributed to Irving Janis (1972) who used it to describe the tendency in a group to reinforce one another's convictions of the appropriateness of their actions and to let 'pressure to conform suppress self-criticism within the group' (Dörner, 1996, p. 34). This means that differing opinions and perceptions are inhibited and not articulated. Groupthink thrives on similarity of group members. On the contrary, while team psychological safety may lead to group cohesiveness, it does so by creating an atmosphere within which the ability to speak up is fostered. Team psychological safety, unlike groupthink, allows opinions that deviate from what the majority in a team think, and to have that opinion respectfully considered, whilst fostering a sense of belongingness and contribution to the team (Tynan, 2005). It is therefore an index of positive use of diversity in a team and implies the tolerating and optimum use of different perspectives, in this case as pertains to risk. There is empirical evidence indicating that reduced interpersonal risk-taking (reduced team/dyadic psychological safety) leads to increased physical risks in certain high-risk industries (Edmondson, 2003), because necessary information may be withheld, distorted or delayed or because of a lack of engagement/ commitment on the part of team members.

It is important to bear in mind that not all potential effects of teamwork are beneficial. The literature shows that increased teamwork has a tendency to reduce individual feelings of accountability (see Muller and Ornstein, 2007, p. 649 for a medical example). In social psychology this has been called *diffusion of responsibility* or what may be called *the snowflake syndrome*.[23] An avalanche at a relatively basic constituent level is just a bunch of tiny snowflakes. However, it is so devastating in its effects that it is difficult to correlate/attribute the consequences of an avalanche to the individual action of any single snowflake. The snowflake syndrome in teams is aggravated by groupthink. Optimum team psychological safety creates the forum within which the team takes ownership of their ideas and are able to present them and have the merits/demerits debated without rancour. This potentially improves team commitment and accountability.

Furthermore, actions to improve interpersonal risk-taking will positively affect the prevention of or recovery from unexpected deviations from quality processes (Deming, 2000) depending on how the information thus gained is fed into organizational learning and action processes. The literature points to one key part of interpersonal risk-taking, that is, the handling of conflict. There is a strong tendency for team members to avoid conflict that in their opinion will be seen by other team members as disruptive. This tendency depends on, among others, such variables as face-giving or threat-sensitivity (Tynan, 2005). Some suggest that the deleterious results of affective conflict (such as high threat-sensitivity and low/ high face-giving) are best mitigated by the stimulation of cognitive conflict:

> Cognitive conflict is disagreement with ideas and approaches. Issues are separated from people. Cognitive conflict is a characteristic of high-performance groups. Affective conflict is interpersonal, with either person-to-person or group-to-group antagonism. Affective conflicts sap energy, sidetrack tasks, and block work. (Garmston, 2005, p. 65)

In support of this Amason et al. (1995, p. 21) indicate that 'conflict is central to team effectiveness because conflict is a natural part of the process that makes team decision-making so effective in the first place'.

This would seem to imply that a key part of the development and training of teams should be the infusion of the understanding that cognitive conflict and questioning should be welcome with due regard to the necessary hierarchical

---

23   'No snowflake in an avalanche ever feels responsible' is a popular quote with attributions to different sources.

structures (see Cosier and Schwenk, 1990 for an examination of this principle as applied to corporate management).

A number of theories suggest ways in which social/organizational dynamics such as conflict may affect team/dyadic psychological safety. Two such theories – Social Exchange Theory and Organizational Support Theory – are particularly pertinent.

Social Exchange Theory

Social exchange theory is based on the thinking that humans behave in a manner that is not unlike that found in economic transactions (motivated by rewards and costs – actual or expected) even when acting in non-economic social settings (Blau, 1964; Homans, 1958; Thibaut and Kelley, 1959). Research supporting this theory suggests that 'social behaviour is an exchange of goods, material goods but also non-material ones, such as the symbols of approval or prestige' (Homans, 1958, p. 606), what Emerson (1976, p. 336) describes, for simplicity as 'the economic analysis of non-economic social situations'.

According to Meeker (1971, p. 485):

> The basic assumption here is that human social behavior can be logically derived or predicted from premises held by the actor whose behavior is being predicted. These premises include (1) his values, (2) his perception of the alternative behaviors available to him, (3) his expectations of the consequences of these alternatives to himself and others (consequences including the probable responses of others), and (4) a 'decision rule,' which is a kind of social norm telling him how the first three premises should be combined to yield a prescription for his behaviour.

It does not seem accurate to imply that 'social exchange theory' is exactly analogous to economic transactions (Zafirovski, 2005) or that it explains all of human motivation and behaviour. The 'theory' does not present the whole picture, but it does present part of the picture and gives a useful reference framework for studying the actions of individuals in dyads, groups and organizational settings. One would add that exchange dynamics are apparent even when rewards come – not directly from the individual recipient of one's actions – but from the larger societal context or environment as well. In other words, the consideration of an exchange is not limited to other individual persons but to larger societal entities. This establishes the underlying figure of exchange as applicable in the context of wider social networks (Emerson, 1976; Yamagishi, Gillmore and Cook, 1988). Thus the rewarding nature of a behaviour (action or inaction) need not be attributed to other individuals in say, a dyad or small group (a micro context), but may be sourced from a larger (macro) context, for example an organization. This type of exchange Emerson (1976, p. 357) calls 'productive exchange' and notes that 'unlike the direct transfer of valued items in simple exchange, here items of value are produced through a value-adding social process'. This is an important point in the consideration of social exchange in an organizational context.

'Social behaviour as exchange' need not be considered a theory, but as 'a frame of reference within which many theories – some micro and some more macro – can speak to one another, whether in argument or in mutual support' (Emerson, 1976, p. 336). In a world where perfect free choice is limited, this framework may also be said to be mediated by other 'rules' – of altruism, competition/rivalry, reciprocity, equity and status consistency/rank equilibration – and 'be subject to processes such as imitation, pressures to conform to norms being followed by others, and degree of commitment to the exchange' and to expectations of what is appropriate in the formal context (Meeker, 1971, pp. 487,492). Furthermore it is appropriate to extrapolate the original theorizing about 'social exchange' – which had a bias towards cognition and rationality (Homans, 1958) – by including an affective element (Lawler, 2001; Lawler and Thye, 1999). This extrapolation leads to the conclusion that, in addition to the external existence or expectation of rewards/costs, the productive exchange process is governed to a large extent by an individual's affective dispositions which act as internal reinforcements or *demotivators*.

The framework of social exchange considered is consequently reliant not only on reinforcement of actions à la Homans or on the a priori consideration of rewards based on rationality à la Blau but also the incorporation of emotions 'as an explicit, central feature of social exchange processes' (Lawler, 2001, p. 321). Lawler (2001, p. 325) defines emotions as 'positive or negative evaluative states with physiological and cognitive components' and notes that they may be 'transitory or enduring, objectless or object focused, and of varying intensity'. He further distinguishes between emotion and sentiment considering the former as 'global or specific, transitory feelings – positive or negative – that constitute an internal response to an event or object' and the latter as 'enduring affective states or feelings about one or more social objects (like relations, groups, self and other)'. Sentiments link emotions (feelings) to social units:

> Structural interdependencies among actors produce joint activities that, in turn, generate positive or negative emotions; these emotions are attributed to social units (relationships, networks, groups) under certain conditions, thereby producing stronger or weaker individual-to-collective ties; and the strength of those group ties determines collectively oriented behaviour, such as providing unilateral benefits, expanding areas of collaboration, forgiving periodic opportunism, and staying in the relationship despite alternatives. (Lawler, 2001, p. 323)

It is also deemed relevant to recognize that there are 'cross-cultural variations in the norms and rules that regulate social exchange' (Cook, 2000, p. 688) and that these must be duly considered in any examination of cross-cultural functioning units.

One key area of exchange is that between a leader and members of a team. In light of Edmondson's (1999) observation that one of the critical drivers of team psychological safety is team leader behaviour, the dynamics of leader-member exchange, especially 'leader inclusiveness' becomes a particularly important area of study. Nembhard and Edmondson (2006, p. 947) propose the construct of leader inclusiveness and define it as 'words and deeds by a leader or leaders that indicate an invitation and appreciation for others' contributions ... [The construct] captures attempts by leaders to include others in discussions and decisions in which their voices and perspectives might otherwise be absent'.

This critical construct has implications for a team member's sense of organizational support, based on the perception that a leader represents the organization. Leadership is an important agency for organizational learning (Child and Heavens, 2001; Sadler, 2001), being instrumental in establishing (or inhibiting) a culture that optimizes learning processes. According to Hofmann, Morgeson and Gerras (2003), research relating to leader-member-exchange (based on role and social exchange theories) lead them to believe that in high quality relationships, subordinates reciprocate the inclusive actions of leadership by enlarging their roles beyond normal role requirements. They believe that 'the type of behaviour valued in the work environment will provide the direction for subordinate reciprocation' (p. 171). What this suggests is that organizational behaviour which is characterized by 'leader-inclusiveness' will help create a climate within which worker engagement for the achievement of quality objectives is enhanced together with increased team psychological safety. 'Leader' in this context should not be considered to be restricted to the on-the-scene supervisor (for example, the master) but also the leadership support from the shore based organization.

Organizational Support Theory

Organizational support theory research literature (for example Eisenberger et al., 1986; Rhoades and Eisenberger, 2002) indicates that employees develop a sense of identity and affective commitment with the organization in which they are employed in keeping with their perceptions of the organization's support and commitment to them. Individual employees ascribe anthropomorphic attributes to the organization based on the actions of its agents and manifest behaviour along the lines of exchange theory as outlined earlier. Social exchange in this context is quasi-dyadic since the anthropomorphized organization is treated as a single entity by the employee. Depending on whether such anthropomorphic actions are perceived to be supportive or malevolent, the individual may respond with increased commitment to the organization and identification with its goals or on the other hand with lower performance that can be gotten away with or in the extreme case even sabotage. Apart from individual values and dispositions, the existence and application of this norm of reciprocity is also based on the individual's cultural exposure and the importance that the particular culture attaches to social exchange as opposed to pure altruistic motivation.

Increased affective attachment resulting from perceived organizational support is thus beneficial to an organization. Eisenberger et al. (1986, p. 506) note that such attachment increases 'the tendency to interpret the organization's gains and losses as one's own ... and the internalization of the organization's values and norms' and to reciprocate with positive action with respect to organizational goals.

A couple of classic motivational theories may be said to augment organizational support theory. They include the expectancy (Porter and Lawler, 1968; Vroom, 1964) and equity (Adams, 1965) theories. The former postulates that employees' task motivation derives from their expectation of a given outcome (instrumentality) and the attractiveness of the outcome to the individual (valence). The equity theory suggests that an employee compares his/her inputs with respect to task performance with perceived outcomes and that if there is incongruity between the two that individual will act to correct the inequity. In arriving at this sense of equity/inequity, this comparison is not limited to the specific individual's input-outcome ratio, but also to other employees' as well.

Organizational Behaviour and Global Shipping Structures

The foregoing discussion draws from long-standing theories regarding dynamics in social settings as well as individual motivational issues. They serve as an important theoretical backdrop for the issue of concern in this study, that is, how shipping companies as organizations learn from, filter and give credence/ acceptability to differing risk perceptions and how this influences the work culture with special regard to group/team dynamics and individual motivation.

Aldrich defines organizations as 'goal-directed, boundary maintaining, and socially constructed systems of human activity' (1979) and maintains that they are important because they are the 'fundamental building blocks of modern societies and the basic vehicles through which collective action occurs ... and mediate the influence of individuals on the larger society' (1999, p. 5). The increase in the amount of research into organizational theory and the role of individuals as well as structure in organizations (Mullins, 2005) is a direct result of this acknowledged importance of organizations.

Studies into organizational behaviour have traditionally been closely related to 'management' which has beginnings in the early twentieth century when the concept of 'scientific management' had found a place in newly industrialized America and Europe.[24] As the century progressed there was an increasing focus on factors that dealt with 'human relations'. The motivation of workers became a

---

24   Max Weber, Frederick Taylor, Henri Fayol, Henry Gantt, Frank and Lillian Gilbreth and Mary Parker Follet, among others, developed concepts of management that either emphasized the scientific, mechanical nature of work and a hierarchical organizational structure or the importance of social as well as economic influences on human behaviour and output.

key consideration of organizational studies. Kanter (1993, p. 23) notes, regarding this 'human relations school' and with reference to Roethlisberger and Dickson[25] that 'people are motivated by social as well as economic rewards and that their behaviour and attitudes were a function of group memberships'. Further this model 'emphasised the roles of participation, communication patterns, and the leadership style affecting organizational outcomes'.

This focus on the systematic study of the attitudes and behaviour of individuals and groups and their impact on organizational effectiveness, led to a quasi-new discipline called 'organizational behaviour'. Organizational behaviour has been defined as 'the study of the relationship between the behaviour of people in organizations and organizational, individual and social outcomes' and helps 'to make sense of a range of activities from the mundane to the critical' (Mills, 2007a, pp. 13,17). The internal processes of organizations are in view here. This relatively narrow view is not sufficient for describing the complexity that all agree characterizes organizational behaviour (Morgan, 2006, p. 345). As Argyris puts it:

> Organizations are extremely complex systems. As one observes them they seem to be composed of human activities on many different levels of analysis. Personalities, small groups, intergroups, norms, values, attitudes all seem to exist in an extremely complex multidimensional pattern. The complexity seems at times almost beyond comprehension. (Argyris, 1990, p. 11)

Definitions of organizational behaviour generally identify the themes of individual and group behaviour, structural patterns and the effect of these on organizational performance and effectiveness (Mullins, 2005, p. 26). It may be added that at both individual and group levels, organizational behaviour is a reciprocal relationship with the external (F.M. Wilson, 2004). Societal paradigms and discourses such as capitalism, socialism, post-colonialism, globalization, the information technology revolution, global security, racial/gender equity and diversity management significantly affect the development of organizational thinking and behaviour. Both inter and intra behavioural patterns (with respect to the individual, subgroup, group and organization) reflect the global and personal discourses in existence, as such making culture a key consideration in organizational studies. Such patterns are also influenced by any particular organization in how it adapts, reflects or utilizes the existing discourse and may in turn influence actions and perceptions on a wider, more global socio-cultural scale. It is interesting that two scholars who represent the extremes of views regarding capitalism (which has very much affected how organizations behave globally) – Adam Smith and Karl Marx – both agree that 'the job makes the person' (Kanter, 1993, p. 3): 'But the understandings of the greater part of men are necessarily formed by their ordinary employments' (Smith,

---

25   Fritz Roethlisberger and William Dickson together with Elton Mayo were involved in research to find work conditions under which productivity was optimized – the Hawthorne Studies.

1976, p. 302); 'It is not the consciousness of men that determines their existence, but, on the contrary, their social existence determines their consciousness' (Marx, 1904, pp. 11–12).

Depending on one's objectives, intentions and perceptions, organizational behaviour can be further defined/analysed in terms of managerialist, actionalist, feminist, racio-ethnic, radical or postmodernist (Mills, 2007a). The managerialist/ actionalist perspective is most suited for the purposes of this work. Of this Smircich (1983, p. 351) – without using the same terminology – notes that:

> The focus of this form of organizational analysis is on how individuals interpret and understand their experience and how these interpretations and understandings relate to action … The researcher seeks to examine the basic processes by which groups of people come to share interpretations and meanings for experience that allow for the possibility of organized activity. The research agenda here is to document the creation and maintenance of organization through symbolic action … *Theorists and practitioners alike are concerned with such practical matters as how to create and maintain a sense of organization and how to achieve common interpretations of situations so that coordinated action is possible.* [Emphasis added]

In general, however, Mills (2007a, p. 37) offers a definition of organizational behaviour which encompasses the various perspectives and is quite comprehensive: 'the study of the impact of behaviour in organizations on organizational, individual and social outcomes'. This is the definition in mind in this study. Aldrich (1999, p. 1) bemoans what he perceives to be an undue emphasis on structure and stability at the expense of 'emergence and change' when research has been done on organizations. In a parallel manner, this research attempts to examine not just organizational structures and stability of organizational behaviour, but importantly the emergence of organizational behaviours, the processes of change as regards the management of risk and learning and the achieving of optimum performance in the maritime industry considering the role of individuals and teams.

Shipping, like other industries, has evolved in the milieu of global organizational behavioural change over the years. Being a derived demand of trade, it has been affected by global trade, economic and societal concerns. However, it seems that the shipping industry – exhibiting a truly 'free market' nature – has survived various global discourses and emerged with a set of unique patterns of organizational behaviour, goals and structure that bridges the diversity of all the players in the field.

It can be argued that, in one sense, the shipping industry is the most global or international of industries, being characterized by an impressive adaptability to the changing discourses of the decades. However, in another sense, it is the most conservative of industries, retaining organizational structures and goals (especially on board ship) that can be argued to be anachronistic.

The setting of the maritime industry in a formal organizational structure has its origins in classical antiquity (Meijer, 1986). There has been subsequently a seemingly irreversible trend toward formalized organizing from building, buying, operating, and selling to scrapping ships. All sections of this process are governed one way or the other by some form of organization. This is symptomatic of human life in the twentieth century and after. It is a life highly dependent on and regulated by organizations – from birth to death. In the process, humanity has come to expect a certain type of behaviour from organizations be they religious, regulatory, administrative or political. These expectations exist for shipping too. They include safe, secure and environmentally friendly shipping. To some extent the industry has also played a role in shaping the global public's expectations. The increased efficiency and cost-effectiveness of shipping has led to growing expectations of better and better service at lower and lower costs. Similarly the consequences of major maritime accidents, especially where they have involved oil spillage, have led to greater expectations of industry accountability. In summary, the expectations are for excellence and quality – irrespective of how groups of people may define these goals. These expectations are accentuated by the current socio-industrial discourses of globalization, climate issues, the management of diversity, the information technology revolution, the issues of global security and relatively recently the 'global financial crisis'.

Despite this obvious importance of the domain, there is a perception of a lack of research into organizational behaviour that considers the specific context of safety in the operation of ships. The uniqueness of the shipping industry – its global expanse and the tendency for diversity in labour force – would seem to suggest that this was a fertile field for such research and its application. To the contrary, organizational behaviour in shipping has, generally speaking, been dependent on long-held tradition. In this area – as in others – the industry shows a high level of conservatism. In a Lloyd's List article it was noted that 'the shipping industry is sometimes very reluctant to act on evidence which seems to fly in the face of its custom and practice' ('Inertia fatigue,' 2007, p. 7).

Such notions as reengineering, learning organizations, and so on, seem to have passed the industry by, at least when it comes to documented research and development. Without categorically affirming that any of these trends are desirable, what the situation suggests is that the organizational structures and practices of the shipping industry are so ideal and perfect as to preclude any tweaking or that shipping organizations are quite isolated from more mainstream organizations ashore. This may be the right strategy for organizational effectiveness, being a kind of 'strategy of avoidance' (Mills, 2007b, pp. 61–2), where the temptation to adopt practices just on the basis of their popularity is resisted. Indeed it may be an asset for any organization to recognize the failings of any such management idea/trend and create interventions that meet the needs of the particular organization and not necessarily in response to the currency of the idea (Abrahamson, 1996). Resistance to certain forms of change (especially when these changes are superficially associated with the current management fad) may lead to organizational survival in a highly competitive business world.[26] However, this state of inertia that characterizes the maritime industry may also be due to a lethargy integral to the industry's social structures and which influences agents/individuals. These agents/individuals, in turn perpetuate the condition by recursive action and thought patterns which may be argued to have negative consequences on the potential effectiveness of shipping organizations as far as safety and environmental protection are concerned. The norms and practices of the industry in the last century have been driven by regulation; the status quo is maintained when the reluctance/lethargy is not directly confronted by new legislation/regulations.

Structure and Agency in a Shipping Organizational Context

With the preceding in mind, the question is raised as to what is the best option for use as a unit of analysis in organizational studies. In examining the dynamics of groups/dyads/organizations regarding risk and its perception and construal, is the best unit of analysis the individual, the team or the organization as a social unit? The approach and methodology of the research is based on the discussion of structure and agency that follows. As indicated by the arguments of Giddens (1984), the focus of study should be the recursive social practices across time and space as tempered by societal structures. Specific studies are limited to the behaviour at the level under consideration and should necessarily entail the examination of how behaviours and attitudes are affected by interactions with the lower (narrower) and upper (wider) levels. The current research approach is informed by 'Open System Theory' (Katz and Kahn, 1978) and Theory of Structuration (Giddens, 1984) as macro-theoretical fundamentals. The open system approach is used because it is considered that the challenges in the maritime industry are best approached holistically. The approach recognizes that the maritime industry is affected by a constant flux in dynamics both inherent in itself and in the external environment and that human effort and motivation are the most important source of maintaining social systems in optimum states (Argyris, 1990; Katz and Kahn, 1978, p. 3). Two things mark this approach:

1. The view that organizations do not exist in isolation, but are open systems in interaction with their environment, with cycles of input, throughput and output (French, Kast and Rosenzweig, 1985; Katz and Kahn, 1978).
2. The consequential need of constant study and learning in recognition of the contingent character of social systems since organizations are constantly affected (at multiple interrelated levels) by a dynamic external environment (Katz and Kahn, 1978).

---

26   See an interesting discussion of this by Robert Bacal at http://www.work911. com/articles/mgmtfad.htm retrieved 22 January 2009.

The studying of the global research question – how shipping companies as organizations learn from, filter and give credence/acceptability to differing risk perceptions and how this influences the work culture with special regard to group/ team dynamics and individual motivation – warrants an examination of factors that are both individual and organizational (micro and macro level social analysis) to which open systems theory, as articulated by Katz and Kahn, is deemed relevant. In their words:

> Open systems theory ... emphasizes, through the basic assumption of entropy, the necessary dependence of any organization upon its environment. The open system concepts of energic input and maintenance point to the motives and behavior of the individuals who are the carriers of energic input for human organizations; the concept of output and its necessary absorption by the larger environment also links the micro and macro levels of discourse ... [And thus] can provide a comprehensive framework for bringing together the advances in organizational research which in themselves are limited and incomplete. (Katz and Kahn, 1978, pp. 15–16)

It thus moves away from the structural functionalist approach to studying social issues which deemphasized the study of the individual and his/her relationship with the environment at large in favour of the macro structures and functions of a societal whole. The individual is deemed a key consideration in societal analysis but is still set in the context of the macro social structure and the external environment. Such thinking is further elucidated and taken one step further by Giddens (1984) – in his 'theory of structuration' – who notes, for example that:

> If interpretive sociologies[27] are founded, as it were, upon an imperialism of the subject, functionalism and structuralism[28] propose an imperialism of the social object. One of my principal ambitions in the formulation of structuration theory is to put an end to each of these empire-building endeavours. The basic domain of study of the social sciences, according to the theory of structuration, is neither the experience of the individual actor, nor the existence of any form of societal totality, but social practices ordered across space and time. (1984, p. 2)

Social theorists have traditionally resided in two camps: one that considers social structure and macro-societal elements as the unit of social analysis and another that

---

27   A reference to the sociological traditions which emphasize hermeneutics and individual subjectivity in the social construction of reality.

28   A reference to the sociological traditions of structural functionalism which emphasize the existence of systems in society (structures) which meet specific societal needs (functions) and constrain individuals and groups to specific behaviours. Social analysis is therefore seen as a more objective analysis of these structures and their functions as opposed to the meanings and subjectivity derived from specific individuals or groups.

sees the individual as the unit of such analysis. According to Giddens (1984) this is an untenable situation. Social analysis should concern itself neither solely with the macro society as structure nor exclusively with the individual as agent, but with a pragmatic blend of the two, with societal systems being generated by the recursive spatial and temporal discourse of human agents. In other words, the recursive actions of individuals in time and space, as shaped by organizational structures and contexts, significantly influence any specific individual, who in turn as an agent influences the organization. Organizations shape individual behaviour not by omnipotent or irresistible control, but by that social constraint that is tempered by the individual's own power in the context (Giddens, 1984, p. 17).

The focus of analysis therefore, becomes the domain of social practices ordered across space/time, keeping in mind the 'duality of structure' and the 'conditions governing the continuity or transformation of structures, and therefore the reproduction of social systems', what Giddens calls 'structuration' (p. 25). This is the approach utilized in this context. The analysis is conducted within the framework that considers the social practices of shipping organizations and their seafarers as a whole mediated by organizational learning influences over time and space and discusses the social systems that have evolved (or are liable to evolve) as a consequence.

Chapter Summary

The chapter covered some long-standing discussions about risk and its construal in social settings, noting that despite the ongoing debate about the subjectivity and/or objectivity of risk, both components were necessary considerations in a pragmatic approach to risk governance, one as an ontological reality and the other as an epistemological hierarchy.

The Social Amplification of Risk Framework (SARF) is therefore theoretically defensible as an important reference for studying the influence of social filters on the communication of risk and by extension, risk governance. The SARF subsumes a number of psychological, social, institutional and cultural processes which create conditions leading to the 'amplenuation' of risk signals. These processes are essentially based on qualitative dimensions that characterize the psychometric paradigm of risk and may influence rational actor risk choices and communication across social strata.

Team/dyad psychological safety is recognized as one of the social dynamics involved in the perception and communication of risk and which leads to increased/decreased interpersonal risk-taking and subsequently affects 'real risk'. Other relevant theories that influence such dynamics are social exchange and organizational support theories. The two create frames of reference from which the conditions that lead to or detract from team psychological safety can be studied.

Organizational behaviour is complex and especially so in the shipping industry. Its study requires a multidisciplinary approach which addresses behaviour and motivation at many levels. In researching this it is important to clarify the units of analysis. An 'open systems' approach helps to acknowledge the diverse influences of external and internal factors on organizational functioning whiles Gidden's theory of structuration bridges the 'unhealthy' divide between the extreme structural functionalist approaches and the hermeneutical/subjective focus on the individual. It makes the unit of analysis the recursive practices of individuals as agents in a structured organizational setting and across time and space.

The chapter serves, not to find gaps in the theory per se, but to note divergences of views and to present the theoretical bases for the research. It points to the paucity of theory-based research specific to the maritime context/literature in this domain and aims, as a contribution to the literature, to address this gap. The essential import of this chapter is that safety and risk are social constructs as far as epistemology goes. It is therefore necessary that systems that seek to optimize risk assessment and management processes prioritize communication and learning in seeking to test the 'reality' of these individual and collective constructs.

Chapter 3
Organizational Culture and Learning

Learning is not compulsory ... but then neither is survival.

W. Edwards Deming[1]

Purpose and Outline

This chapter is the second of two setting out the theoretical foundations for the research. Concepts discussed include culture, organizational learning theory, worker engagement and motivation for learning. The discussions revolve round issues of definitions, types, levels and paradoxes.

Culture and Climate

The nature of individual, group and organizational behaviour and the underpinning influences of social structures, as discussed in Chapter 2, make culture a significant consideration in an analysis of organizational effectiveness relevant to the research focus.

Like many of the concepts that relate to characteristics of human communities, there appears to be no universally acclaimed definition of culture. The difference in the definitions of culture probably lies, to some extent, in the fact that the construct is one associated with collective human behaviour. The levels of aggregation of these 'human collectives' determine not only the field of study, but also how culture is defined and what 'dimensions' constitute it. In what Hofstede calls a 'well-known anthropological consensus definition', Kluckhohn (1951, p. 86 as cited by Hofstede, 2001) defines culture as consisting:

> in patterned ways of thinking, feeling and reacting, acquired and transmitted mainly by symbols, constituting the distinctive achievements of human groups, including their embodiments in artifacts; the essential core of culture consists of traditional (that is, historically derived and selected) ideas and especially their attached values.

At the organizational level of aggregation culture has been defined as:

> A pattern of shared basic assumptions that was learned by a group as it solved its problems of external adaptation and internal integration, that has worked well

---

1   Retrieved 5 May 2009 from http://www.iwise.com/0A7fv.

enough to be considered valid and, therefore, to be taught to new members as the correct way you perceive, think and feel in relation to those problems. (Schein, 2004, p. 17)

There is also evidence for 'cultural' differences derived from roles and identities such as profession/occupation, gender, age and/or rank (see for example Bloor and Dawson, 1994, p. 281). Individuals, with their unique personalities, are impacted/ influenced by these 'social exposures' – which are not necessarily deterministic – and subsequently develop behavioural patterns that reflect these influences to different degrees.

Hofstede (2006) gives what he calls a 'shorthand definition of culture' as 'the collective programming of the mind that distinguishes the members of one group or category of people from others'. In his work relating to global work-place cultures he analysed the data from a questionnaire for IBM employees in 72 countries – 116,000 questionnaires administered in 72 countries in 20 languages between the years 1967 and 1973 (Hofstede, 2001, p. 41). From this analysis the argument is made for the existence of relatively stable national cultures, characterized by different dimensions which vary from country to country. This work has been seminal in many ways, generating significant supporting citations, secondary research, critical reviews and disagreements.

Other research efforts (Hall, 1976; House, Hanges, Javidan, Dorfman and Gupta, 2004; Schwartz, 1994; Trompenaars and Hampden-Turner, 1997) agree on the existence of 'dimensions' of national culture and their variability in different societies, without necessarily agreeing as to what these dimensions are. The work of Inglehart and others in the 'World Values Survey' also implies this.[2] However, the entire notion of the existence of cultures specific to nations and the further 'dimensionalizing' of these is severely criticized by other researchers who fault the study on many levels including sampling methodology, construct and internal validity and ecological fallacy (McSweeney, 2002a, 2002b), misapplication of domain definitions (Baskerville, 2003) and western bias and marginalization of the effects of globalization (Magala, 2004) among others. McSweeney even considers the option of dismissing Hofstede's notion of culture, values and dimensions of culture as 'a misguided attempt to measure the unmeasurable' (2002a, p. 90). A particularly worrying consequence of Hofstede's work has been its unquestioned application (30+ years after the sample was taken) to various contexts in a way which views culture as completely deterministic and static. The criticisms have elicited some responses (Hofstede, 2002, 2003), and Williamson (2002) puts forward a substantially credible defence of Hofstede's work while maintaining an objective criticism of some aspects of it. He acknowledges that McSweeney's criticisms of Hofstede have some appropriate warnings for social

---

2   Information regarding the 'World Values Survey' is comprehensively covered on the site http://www.worldvaluessurvey.org/ retrieved 23 January 2009.

and organizational researchers but also points out areas in which McSweeney's criticisms are flawed.

All in all, the literature covering the application of cultural models in research appear to be biased towards the conceptualization of culture as a macro-phenomenon that involves vast societal groupings (such as nations) and tends to see the behaviour of individuals and groups as dominated by this single macro-culture. Contrary to this dominant thinking, Berthoin Antal and Friedman (2005, pp. 73,74) recognize the individual as an entity of 'amazingly complex "cultural composites" '. They cite the work of Swidler who proposes seeing culture as a 'repertoire of capacities from which varying strategies of action may be constructed' (Swidler, 1986 as cited byBerthoin Antal and Friedman, 2005, p. 74). In addition to this complexity, culture may also be selective in its choice of values to champion based on context and timing, leading to the phenomenon of 'value trumping' as described by Osland and Bird (2000). They note with an example that:

> Schemas reflect an underlying hierarchy of cultural values. For example, people working for US managers who have a relaxed and casual style and who openly share information and provide opportunities [for subordinates] to make independent decisions will learn specific scripts for managing in this fashion. The configuration of values embedded in this management style consists of informality, honesty, equality, and individualism. At some point however, these same managers may withhold information about a sensitive personnel situation because privacy, fairness, and legal concerns would trump honesty and equality in this context. This trumping action explains why the constellation of values related to specific schema is hierarchical. (p. 71)

In view of the foregoing and based on a synthesis of the literature (including Hofstede, 2001; Pidgeon, 1997; Spencer-Oatey, 2000; Trompenaars and Hampden-Turner, 1997 and so on), culture is defined in the current study as *'a dynamic[3] intangible and composite system of interacting values, basic assumptions and norms which manifests in and influences individual and group attitudes, beliefs, behavioural patterns and non-behavioural items and which informs the meaning individuals/ groups attribute to such manifestations in themselves and others'*. As Hofstede (1997) and Dahl (2004) suggest, culture is best conceptualized as situated between non-programmable human nature on one side and human personality/individuality on the other hand. The potential effects of culture in creating distinctions between groups of people need not be a definitional issue as Hofstede's (2006) definition implies. Culture may create such distinctions, but in many cases it does not and thus the issue of distinction is not the crux of the **definition** of culture.

---

3   This dynamism includes levels of contingency and emergence – fluid but characterized by significant inertia.

For the purposes of the current study, it is enough to acknowledge the influence of societal norms and the behavioural manifestations of these (as indicated by all of the aforementioned research) and to note that there are similar norms and manifestations in organizations that may shape and influence group and individual behaviour. 'Culture is to an organization what personality is to an individual' (Westrum, 1993, p. 401). Dewey (1930, p. vii) notes that 'custom is essentially a fact of associated living whose force is dominant in forming the habits of individuals'. In acknowledging this generic influence, however, it is important to recognize – in line with the thinking of Berthoin Antal and Friedman (2005) – the significant role of the individual and the multiplicity of cultural influences at play at any one time, simultaneously expanding and limiting the range of behavioural, attitudinal and cognitive responses available to the individual. One basic premise, therefore, of the current work is that the mindset of individuals (influenced and reinforced by relevant societal thinking and behaviour) impact on team psychological safety and that the existence of negative team psychological safety can be mitigated by the dimensions of organizational culture that exist or are brought to bear on group processes. The literature indicates an almost global recognition of the existence of such cultures unique to every organization – corporate cultures – which influence and are in turn influenced by the people who form part of the organization (Peters, 1978; Smircich, 1983).

Recognizing this and furthermore agreeing that culture – whether regarding safety, quality, security and so on – is dynamic and is not 'inheritable or genetic but is learned' (Dahl, 2004, p. 6), gives impetus to a drive to examine the factors that influence these processes in seeking better management, operational and learning practices in the shipping industry. The individual and team level of organizations is critical to this. Culture does not exist as a 'singular, uniform and monopolistic' entity in societies. Rather, a multiplicity of 'dissenting, emergent, organic, counter, plural, resisting, incomplete, contradictory cultures is recognisable in any organization' (McSweeney, 2002b, p. 96; Smircich, 1983). The acknowledgement of the existence of a 'dominant culture' that can be made to positively affect corporate behaviour is therefore not a denial of the presence of other 'cultures and subcultures'.

As part of this theoretical review, there is the need to clarify difficulties relating to the differences between organizational 'climate' and 'culture'. The two terms have often been used interchangeably. However, the literature suggests a distinction between them, while still retaining considerable discussion as to the differences between culture and climate as well as the possibility of generating dimensions and levels in both (Denison, 1996; Ek, 2006; Guldenmund, 2000; Pettigrew, 1979; Reichers and Schneider, 1990). This author takes the position that climate is the manifestation of culture in a relatively short time frame – a snapshot, as it were, of a much more complex but latent and dynamic construct (culture). Both concepts represent the social contexts of life in organizations, but address different temporal scopes and sensory manifestations. In other words climate can be assessed in a short frame of time and with more superficial observations, while culture is more tacit with respect to the latter and more long-term with respect to the former. The evolution of culture is the causative agent of climate. Organizational 'climate refers to what happens in an organization and culture refers to why it happens' (Ostroff et al., 2003 as cited by Carr, Schmidt, Ford and DeShon, 2003, p. 614). This is in agreement with the views of Guldenmund (2000, p. 221) who concludes from his review of the literature that the term 'organizational climate' initially signified a broad construct for researchers but over time has been seen more as defining 'organizational culture' with climate taking on the more restricted meaning of 'attitudinal or 'psychological' phenomena within an organization … the overt manifestation of culture within an organization'. This view makes climate more observable and 'recordable' (thus lending itself to more quantitative modes of research) and culture more subjective and implicit (lending itself to more qualitative methods of research).

In matters associated with risk/safety, one of the key elements that may be argued to show an organization's culture and limitations is learning (Pidgeon, 1997; Reason, 1997). The literature on learning processes and outcomes in organizations is discussed next.

Organizational Learning and Learning Organizations

Do organizations learn? Can organizations learn? 'What is an organization that it may learn?' is an appropriate question posed by Argyris and Schön (1996, p. xx).

They see organizations as 'behavioural settings for human interaction, fields for the exercise of power, systems of institutionalized incentives that govern individual behaviour, or socio-cultural contexts in which individuals engage in symbolic interaction' (Argyris and Schön, 1996, p. 7). It is possible to visualize organizations as represented by a number of metaphors: machines, organisms, brains, cultures, political systems, psychic prisons, flux and transformation, instruments of domination (Morgan, 2006), as a spider in its web (Wheatley, 1999) and so on. As Morgan notes (pp. 337–43), organizations are complex and these metaphors are not static, being time and context dependent as well as having limitations in their representation of the whole and of reality. A particular organization may exhibit a number of them at the same time or progressively. Some of the metaphors (for example, as organisms, brains, cultures, flux and transformation, and so on) make organizational learning a distinctly valid conceptualization for optimizing organizational performance.

The currency of organizational learning (present since Weber)[4] is acknowledged in the literature (see for example Argote, McEvily and Reagans, 2003). Its modern relevance is emphasized by the issues raised in comprehensive

---

4   Max Weber's early 20th century work on bureaucratic forms of organizations had clear connotations of organizational learning issues; rules had to be learned and the organization had to have systems of recording and storage.

interdisciplinary handbooks on the topic in 2001, 2005 and 2006 (Dierkes, Antal, Child and Nonaka, 2001; Easterby-Smith, Araujo and Burgoyne, 2006; Easterby-Smith and Lyles, 2005). These references raise and discuss the contemporary need for enhanced organizational learning due to the importance of such issues as globalization, transfer of technology, human resource and markets across borders, rapid change in an information age and national/international security as well as the continued emergence of a 'risk society' (Beck, 1992; Mythen, 2004). Whenever there is a requirement for change in any industry, learning and how it takes place in organizations, becomes an issue for examination. The result has been the proliferation of many conceptions of organizational learning as addressed by different academic disciplines – from economics to psychology (Dodgson, 1993, p. 375).[5]

Despite this agreement in the literature about the necessity of organizational learning and the validity of organizational learning as a research construct, there is some inter and even intra-discipline ambiguity in what specifically constitutes the construct under investigation (Dodgson, 1993; Fiol and Lyles, 1985). This ambiguity is also present in the debate as to whether organizational learning is more cognitive than behavioural, or indeed whether the construct speaks more to processes or products/outcomes. In a review of the pre-1985 literature on organizational learning, Fiol and Lyles (1985, p. 811) separate learning from adaptation: learning is 'the development of insights, knowledge and associations between past actions, the effectiveness of those actions, and future actions' while adaptation is 'the ability to make incremental adjustments as a result of environmental changes, goal structure changes, other changes'. In other words, learning is not adaptation but a precursor to adaptation/change in behaviour. Environmental change (conflict between established practice, routines, mindsets, paradigms and the dynamic conditions of factors internal and external to an organization) creates a stimulus for learning, which then – depending on the actors and the level and kind of learning – leads to organizational adaptation, evident in strategies, policies and measurable outcomes. The notion of 'learning' – whether as a process or as an outcome linked to change in behaviour (Argyris and Schön, 1996, p. 3) – is arguably applicable to individuals, teams and organizations. When seen as a process, rather than solely as an outcome, one can look for specific behaviours, whether at the individual, team or organizational levels which would indicate learning – activities and processes that show that the individual, team or organization carries out 'an iterative process of designing, carrying out, reflecting upon, and modifying actions' (Edmondson, 1999, p. 353 citing Dewey, 1922) and characterized by 'learning behaviour' such as 'asking questions, seeking feedback, experimenting, reflecting on results, and discussing errors or unexpected outcomes openly' (p. 353).

---

5   See Appendix 1 (Glossary) for a list of definitions given for 'organizational learning'.

A broad definition of organizational learning would have to include:

1. Learning in a strict sense as the process of acquiring information or knowledge with the potential to affect cognitive, affective or behavioural outcomes of the learning entity.
2. Learning as sense-making and the abstraction of meaning resulting in the possibility of understanding/perceiving reality in a different way.
3. Memory as the process by which information or knowledge and attributed meaning is stored and recalled.
4. Action as the outcome of these processes manifested as strategy, choices and dispositions in terms of the cognitive, affective and behavioural.

It is important to consider the variability of action/behaviour as an outcome of learning. Action here is taken to include choices of continuity and stability. Behavioural change is primarily the empirical basis for *assessing* learning (Argyris and Schön, 1996, pp. 33–4). However, learning should not be *defined* in terms of change in behaviour. Learning may just as well result in no change in behaviour, being essentially a modification in 'cognitive maps or understandings' (Friedlander, 1983, p. 194). While these cognitive maps or understandings may manifest in behaviour, the behaviour itself is only evidence of learning and not learning in essence. Furthermore strategy resulting from learning, could very well lead to a decision for continuity and stability – no change in behaviour. An action need not be taken as a result of learning. When learning occurs, what does change is the *range of potential* behaviours (Huber, 1991). Actual behaviour change then becomes an indicator that learning has taken place, but does not in itself define the learning.

In line with this, the following generic definition of organizational learning is offered:

> The intentional or unintentional processes by which organizational social systems (made up of individuals and groups) create the potential (in cognitive, behavioural and/or affective terms) to adapt in behaviour, values and attitudes as a result of exposure to experiential, vicarious or contextual events whether or not there is a conscious awareness of this process. Learning also includes the process of hypothetically/rationally inferring causal links not explicitly based on historicity but linked to patterns stored in memory.

It is noteworthy that organizational learning does not necessarily lead to improved performance or effectiveness (Huber, 1991, p. 89; Maier, Prange and Rosenstiel, 2001, p. 16; Miller, 1996, p. 486). In other words a resulting change of behaviour may or may not be positive, leading to the qualifying of desired learning by Argyris and Schön as 'productive organizational learning' (1996, p. 18). To quote them:

There are several ways in which instrumental learning may be for ill rather than for good. Some of these are particular to organizational learning; others, applicable to learning by agents of any kind. First the *ends* of action may be reprehensible. The value we attribute to an increase in effectiveness or efficiency depends on how we answer the question, Effectiveness or efficiency for what? And how we evaluate the 'what.' This issue is critically important when the action in question emanates from an organization whose members are eager or unthinking compliant participants. During World War II, Eichman's bureaucracy learned over time to become more efficient at sending its victims to the gas chambers. (Argyris and Schön, 1996, pp. 18–19)

In the shipping world, examples of learning with consequential unproductive ends can be seen in the issue of fatigue. Some years ago this was not an issue, but then came manning[6] rules that allowed specific kinds of ships to be arguably 'under-manned'. More and more companies have learnt to subscribe to jurisdictions where such manning is permitted and now 'there is a disturbing reluctance to 'rock the boat' when it comes to the long hours culture in shipping' despite the potentially catastrophic cost. (Grey, 2009, pp. 4–5)

To some extent then, organizational learning of some sort may be argued to be going on all the time whether to facilitate dramatic change or to maintain the status quo in inertia (Dodgson, 1993, p. 380).

Learning in organizations can take different forms including congenital, experiential, vicarious, contextual (Bresman, 2005) as well as inferential/inductive ways (Holland, Holyoak, Nisbett and Thagard, 1989).

Congenital Learning

Congenital learning is that learning that is associated with the inception of an organization. It is learning that emanates from the external – the experience of others before the inception of the organization (García-Morales, LLoréns-Montes and Verdú-Jover, 2006, p. 524).

Experiential Learning

Experiential learning may be seen as occurring when the processes of learning are triggered by events that lie directly in the learner's contexts. It is internally generated learning being derived from the entity's (individual or organization) own experiences and the consequences of those experiences to that entity. Many subscribe to the notion that experience is the best teacher. As early as 52 BC,

---

6   In gender-sensitive terminology 'crewing' is more appropriate.

Julius Caesar in 'De Bello Civili',[7] claimed experience to be 'the best master in every thing on which the wit of man is employed'. However, this view is not always necessarily supported by research (Deming, 2000, p. 19; Dörner, 1996, p. 170). Repetitive accidents/incidents indicate that not all experience is positively educative. Even if it were granted that 'experience is the best teacher', individuals, teams and organizations are not always efficient learners. Furthermore, as a popular aphorism puts it, 'experience keeps a dear [costly/expensive] school' (Franklin, 1986, p. 28). Experiential learning is arguably a universal form of learning, but it is limited in that the experience that triggers learning, may threaten the progress and even survival of the learning entity. The literature suggests other ways in which learning at individual, team and organizational levels may be augmented.

Vicarious Learning[8]

This kind of learning occurs when a learning entity learns from others outside its own existential context, that is, learning is derived from the experience of other entities (Chew, Leonard-Barton and Bohn, 1991, p. 10; García-Morales et al., 2006, p. 524). It is a particular form of social learning and modelling (Bandura, 1977). This kind of learning requires the learning entity (the learner) to be aware of the behaviour of another entity (the model) and the recognition of the consequences of that behaviour as either desirable or precautionary for the learner. It is based on such factors as the learner's perception of the relevance of the action/consequence, the similarity of contexts, the differences between the particular model and others, and the learning entity's perceived capacity to act in a similar manner (Maier et al., 2001, pp. 19,24). However, vicarious learning also has limitations. As Nathan and Kovoor-Misra (2002, p. 246) indicate, the biggest inhibitor of such learning is that organizations, while desiring to learn from others, may not be as motivated to do so as they would be in the event of a crisis happening to them. They suggest that there is higher motivation for learning from an organization's own experiences than from those of others. Another limitation is the presence of contextual difficulties in the migration of other's experiences into settings that may be unique to the learner.

It can be seen from the above that the foundations of learning lie in the perceptions of the observer and the environmental/societal/contextual influences

---

7   A translated version of this work – Caesar's Commentaries on the Gallic and Civil Wars – can be found at http://etext.virginia.edu/etcbin/toccer-new2?id=CaeComm.sgmandimages=images/modenganddata=/texts/english/modeng/parsedandtag=publicandpart=all retrieved 17 November 2008.

8   The word 'vicarious' means 'performed or achieved by means of another or by one person etc., on behalf of another' (Murray et al., 1991, p. 2232). Strictly speaking, therefore, 'vicarious learning' would imply that the learning is done on one's behalf by another. However, it is used in this context to mean that the experience, and not the learning, is vicarious.

he or she is exposed to. This leads to an important aspect of all learning, that of context.

Contextual Learning

Contextual learning is based on the notion that all learning is apprehended within the subjective influences arising from the settings/contexts in which the learning entity is situated. The learning experiences that may be encountered vicariously must be considered contextually. According to Tyre and von Hippel (1997, p. 71):

> Different organizational settings (1) contain different kinds of clues about the underlying issues, (2) offer different resources for generating and analyzing information, and (3) evoke different assumptions on the part of problem solvers. Consequently, actors frequently must move in an alternating fashion between different organizational settings before they can identify the causal underpinnings of a problem and develop a suitable solution. These findings suggest that traditional, decontextualized theories of adaptive learning and of collaboration could be improved by taking into account that learning occurs through people interacting *in context* – or, more specifically, in multiple contexts.

Inferential/Inductive Learning

This is another mode of learning that is pointed out in the literature. The inferential/inductive aspect of learning (learning not based on the reinforcement of feedback) cannot be ignored considering the contemporary emphasis on and need for organizational learning based on completely new factors such as technology, informatics and globalization. Induction has been taken to 'encompass all inferential processes that expand knowledge in the face of uncertainty' (Holland et al., 1989, p. 1). Such uncertainty and limited feedback results from the presence of events that are unstructured or too distant in time to afford the consistent occurrence/reoccurrence which is the basis of most learning. Inferential/inductive learning involves exploration, experimentation and sometimes unsubstantiated inferences of causality (Maier et al., 2001, pp. 21–2). It is noteworthy that this creates a dilemma for high-risk industries, where there is limited system tolerance for trial and error processes.

The learning processes described above – congenital, experiential, vicarious, contextual and inferential – are not mutually exclusive, but exhibit significant overlaps. However, they help clarify modes of learning, thus allowing for better research focus.

The literature further characterizes some organizations as 'learning organizations'.[9] The general indication seems to be that what makes an

---

9   Definitions given for a 'learning organization' are indicated in Appendix 1 – Glossary.

organization a learning organization is the ability and practice of going beyond the natural learning that is arguably going all the time – whether to productive or negative ends – and attempting to develop higher level, constructive or generative *system* functions which are reflected in the purposeful development of strategies and structures to facilitate and coordinate productive learning (Dodgson, 1993, p. 380). Such organizations appear to view the process of learning (as opposed to the object of learning) not as a destination, but a lifelong (continuous) journey whose rewards are self-inherent (Senge, 2006, p. 10).

In the context of this study, the following definition of a learning organization adapted from that of Mullins (2005, p. 1057) is intended:

> An organization that *encourages* and *facilitates* [emphases added] the learning and development of people [individually and collectively and with respect to the strategic goals of the organization in the context of the external environment] at all levels of the organization, values [and retains] the learning and simultaneously transforms itself.

The degree to which this benefits the individuals and organization is considerably dependent on how the concept is practically interpreted and the degree of commitment to it as part of the organizational culture. One characteristic of this kind of organization is that it promotes the exchange of information and sustains a flexibility that supports the acceptance and sharing of new ideas and visions. Implicit organizational knowledge is intentionally made explicit and the results could be a potential increase in knowledge spread and assimilation throughout the organization. Reflection, among others, is a distinct part of the ethic of such organizations (Redding, 1997, p. 62).

Finger and Brand (1999, p. 136) consider the 'learning organization' to be the ideal 'toward which organizations have to evolve in order to be able to respond to ... various pressures' as contrasted with organizational learning which they see as 'the *activity* and the *process* by which organizations eventually reach the *ideal* [emphases added] of a learning organization'(Finger and Brand, 1999, p. 136).

Although individual learning is critical to this learning, the routines that make organizational learning visible are 'independent of the individual actors who execute them and are capable of surviving considerable turnover in individual actors' (Levitt and March, 1988, p. 320). As is put by Hedberg:

> Although organizational learning occurs through individuals, it would be a mistake to conclude that organizational learning is nothing but the cumulative results of their members' learning. Organizations do not have brains, but they have cognitive systems and memories. As individuals develop their personalities, personal habits, and beliefs over time, organizations develop world views and ideologies. Members come and go, and leadership changes, but organization's memories preserve certain behaviors, mental maps, norms, and values over time. (1981, p. 6 as cited in Fiol and Lyles, 1985)

The learning that an organization achieves is not the sum of individual learning, but includes the synergies of association and social discourse facilitated by these individuals and the systems and structures in place (Child and Heavens, 2001, p. 309). Even where individual learning is significant and positive, social/team dynamics and norms can block organizational learning. Individual learning, however, remains essential for overall organizational learning. In the same way, team learning is of vital importance in organizational contexts. Senge is even of the view that 'team learning is vital because teams, not individuals, are the fundamental learning unit in modern organizations' (2006, p. 10).

This dynamic interplay of agency (human will and behaviour) and social structure, referred to as 'structuration' by Giddens (1984), makes the examination of individual action, teamwork, as well as organizational support context relevant in the study of organizational learning. Organizational learning, as a whole, is generally greater than the sum of learning of its individual members, or of its parts or both. It is a state of gestalt which when actively promoted facilitates even more learning. It is worthy of note that organizational learning may also be less than the sum of learning of its individual members/parts (Argyris and Schön, 1996, p. 6; Hedberg, 1981).

Single-Loop, Double-Loop and Deutero-Learning

According to Argyris and Schön (1996, p. 20) learning takes place in diverse ways. Single-loop learning, double-loop learning and deutero-learning are important variants that they examine. In single-loop learning, individual or organizational learning is facilitated by a questioning of present actions and policies (action strategies) when trying to correct for detected errors. Double-loop learning, on the hand, questions the underlying norms/values, policies and objectives, what Argyris and Schön call 'governing variables'. The basic concepts are captured by others but with different terminology (see for example: low and high level learning (Fiol and Lyles, 1985); survival, adaptive and generative learning (Senge, 2006, p. 14); and tactical and strategic learning (Dodgson, 1991)).

Figure 3.1 illustrates the concept in simple terms.

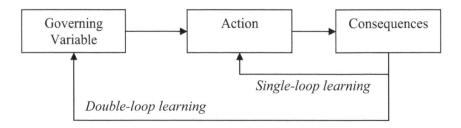

Figure 3.1   Simple   conceptualization   of   single-loop   and   double-loop learning

Unique to Argyris and Schön is the introduction of a special kind of double-loop learning, deutero-learning as applied to organizations.[10] Of deutero-learning they have this to say:

> A critically important kind of organizational double-loop learning ... is the second-order learning through which the members of an organization may discover and modify the learning system that conditions prevailing patterns of organizational inquiry. (Argyris and Schön, 1996, p. 29)

Deutero-learning is premised on an inquiry into inquiry, 'learning how to learn' and characterizes the most exceptional level of learning in organizations.

The double-loop kind of learning (at least) is obviously what should characterize a learning organization as defined above, whereas single-loop learning will be more characteristic of organizations that learn under 'survival' or 'adaptive' conditions.[11] An emphasis on double-loop and deutero-learning requires 'worker engagement' precisely because organizational deutero-learning 'is critically dependent on individual deutero-learning' (Argyris and Schön, 1996, p. 29). This is the case since it is the recurring practices of individuals – as they influence and are influenced by the constraints of the social context of the organization (in line with structuration theory) – that determines the progress of the whole organization in attaining its goals. Worker engagement is therefore hypothesised to be the fore-runner of achieving consistently the organization's quality objectives via productive learning in the maritime industry.

Another insightful perspective into learning is given by Svedung and Rådbo (2006). They define and describe single-loop learning as the process that characterizes dynamic decision-making which is 'forward oriented in time' and is based on tightly coupled (in terms of function and time) experience feedback and 'with no formal methods of data handling and analysis associated with it' (p. 24). This description, while being obviously different from that of Argyris and Schön (1996, pp. 20–25), is a very pertinent and useful view. Svedung and Rådbo contrast their description of single-loop learning with their description of double-loop learning – a process which is 'mainly backward oriented in time and not primarily meant to function dynamically but to supply the actors involved with basic understanding of the system' (p. 24). To avoid the confusion that can result in the ambiguous use of terms, it is suggested that Svedung and Rådbo's concepts retain the label of 'dynamic decision-making' and 'reflexive learning' and that the latter be seen as the essence of Argyris and Schön's definition of both single-loop and double-loop learning. Accordingly, both individuals and organizations may engage in single-loop learning, double-loop learning or both as per Argyris and

---

10   They cite Gregory Bateson (1972) for originating the term 'deuterolearning' as applied to individuals.

11   See Senge's use of the terms 'survival learning', 'adaptive learning' and 'generative learning' (Senge, 2006, p. 14).

Schön, but such learning is essentially reflexive in nature. It is the depth of inquiry that is an issue for Argyris and Schön and their definitions of these two types of learning show limited dynamic decision-making in the sense that Svedung and Rådbo describe.

Conceptual Approaches to Assessing Organizational Learning

The majority of the definitions of organizational learning are not fundamentally different from each other but are complementary. Broadly speaking (and with what is on the surface a process or systems perspective) organizational learning concerns itself with:

1. Knowledge acquisition.
2. Processes that make implicit knowledge explicit.
3. The retention of acquired knowledge as well as knowledge made explicit.
4. The usage of knowledge in strategy and action.

Huber gives a useful classification as shown in Figure 3.2. *Grafting* in Figure 3.2 is basically a reference to mergers and acquisitions.

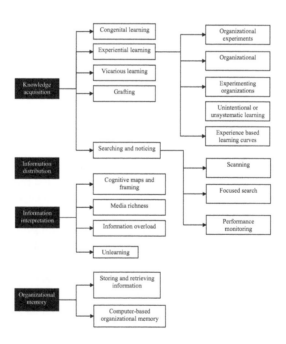

Figure 3.2   Constructs and processes associated with organizational learning

Source: Huber, G.P. (1991, p. 90)

While Huber's approach is often seen as an information processing perspective of organizational learning (Wang and Ahmed, 2002, p. 10), the processes, structures and systems are – as suggested by the structuration theory (Giddens, 1984) – underpinned by individuals and collectives and their recursive practices that become the emergent cultures inhibiting or facilitating organizational learning (Berends, Kees and Weggeman, 2003; Child and Heavens, 2001, p. 322). The recursive practices form the structures identified with 'knowledge management'.

Consequentially, an analysis of the Huber constructs – as pertaining to risk, need not be limited to one perspective of organizational learning,[12] but can conceptually include the theoretical themes of the levels of societal aggregations of learning,[13] types of learning[14] and the role of knowledge management structures (Argote et al., 2003; Templeton, Morris, Snyder and Lewis, 2004). It is important to include all these themes, together with the outcomes of learning (which Huber omits), in order to gain a better understanding of the construct of organizational learning (Nicolini and Meznar, 1995, p. 728) and to 'build on the diversity of perspectives by taking a multidisciplinary approach' (Friedman, Lipshitz and Popper, 2005, p. 27) for pragmatic ends.

Organizational Learning Paradoxes

The essence of organizational learning is to optimize an organization's efficiency through critical inquiry and necessary change. Monitoring this process means that feedback is constantly sought, fostered and acted on. The more positive this feedback is, the more there is a tendency to regard organizational learning as being successful. However, an inherent danger of positive feedback, when deemed to signal organizational success, is that it potentially leads to 'the standardization of action', procedures and structures and ultimately to 'organizational inertia' – the very thing that limits further productive learning (Berthoin Antal, Dierkes, Child and Nonaka, 2001b; Starbuck and Hedberg, 2001). This is a significant paradox in organizational learning. Following Miller (1992), this phenomenon is referred to as the *Icarus paradox*.[15] It is necessary for there to be an appreciation of this tendency and to explore ways in which organizations can merge the merits of

---

12   Wang and Ahmed (2002, pp. 9–12) list the different perspectives as focusing (1) on individuals' influence on collective learning, (2) on processes and systems as an extension of systems thinking and developing from information processing, (3) on knowledge management, and (4) on continuous improvement and incremental innovation.

13   The interaction between individual learning and learning at different levels of aggregation of collectives.

14   The distinct modes of organizational learning, e.g. experiential, vicarious, contextual, inferential learning as expressed in different levels of learning: single-loop, double-loop or deutero-learning.

15   A Greek myth tells the story of how Icarus made wings of feathers held together by wax in order to fly. Ignoring the warnings of his father, he flew higher and higher. However, the success of his endeavours had inherent in itself the potential for failure, in that

continuity with those of necessary change. According to research by Berthoin Antal, Krebsbach-Gnath and Dierkes (2003),[16] it is possible to avoid this and thus to challenge this particular paradox of organizational learning.

Argyris and Schön (1996, p. 281 ff.) describe another learning paradox, one associated with single-loop learning. They describe it as 'the actions we take to promote productive learning but which actually inhibit deeper learning'. It is specifically associated with single-loop learning where the practice of avoiding 'undiscussable' domains and only generating solutions for discussable features of problems, may be present. They explain that over time these practices undermine single-loop learning and inhibit double-loop learning.

The issue of the coexistence of diversity and consensus is yet another paradox in organizational learning. It is best described by Fiol who states that:

> Minds as diverse as John Dewey (1954) and Thomas Kuhn (1970) have argued that all knowledge development rests on unity of interpretations ... Yet learning, or the development of new knowledge, is also based on diversity of interpretations (Bower and Hilgard 1981). To remain viable in the face of change, organizations have to maintain both prescriptions in balance. To learn as a community, organizational members must simultaneously agree and disagree ... Learning in organizations entails not only the acquisition of diverse information, but the ability to share common understanding so as to exploit it. The apparent paradox is that collective learning, by definition, encompasses both divergence and convergence of the meanings that people assign to their surroundings. (Fiol, 1994, p. 404)

Organizational Credence-Giving to Risk Perceptions

An organization is exposed to different sources of risk information. This information is filtered through the perceptions of leaders and subsequently given credence based on often very subjective determinants (Kozlowski and Doherty, 1989). For the productive learning described in earlier sections of this chapter to be thorough and achievable, there must be room for depth of inquiry (double-loop learning) that characterizes the organization as a whole. This implies a pervasive understanding that the risk perceptions of individuals and teams are not only tolerated, but actively sought and negotiated (Edmondson, 1996). Organizations, especially those that are in high-risk industries, must have cultures of credence-

---

the more successful he was (i.e. flying higher) the closer he came to the sun and increasingly had his wax wings melting, ultimately causing him to plunge to his death.

16   The research examined the then German pharmaceutical company – Hoechst – and discusses the effects of radical learning in what was already a very profitable and stable company.

giving to varied perspectives on what constitutes risk and allow dissenting opinions to be heard in a manner that generates cognitive but not affective conflict (Cosier and Schwenk, 1990). The literature suggests that where risk perceptions are stifled or where credence is given to only those sources that have status, based on rank, gender, nationality, time in employment, professional qualifications and so on, the organization closes off potentially valuable sources of risk information (Amason et al., 1995, p. 28; Fiol, 1994). This amplification and attenuation of risk signals based on social dynamics of power, culture and politics limit optimum productive learning processes as discussed earlier.

Credence given to an individual or team's perceptions of risk, influences organizational climate and may increase motivation to learn and worker engagement (Carr et al., 2003). It signals an organization's acknowledgement that productive organizational learning and change 'are not the outcome of heroic leadership, but rather require leadership at many levels' (Berthoin Antal et al., 2003, p. 8). This is an important principle for this study. The value of intrinsic motivation as against extrinsic motivation in individual and organizational behaviour and learning has been well noted (Ryan and Deci, 2000) as has the importance of self-regulation. Figure 3.3 shows such a continuum/taxonomy of motivation.

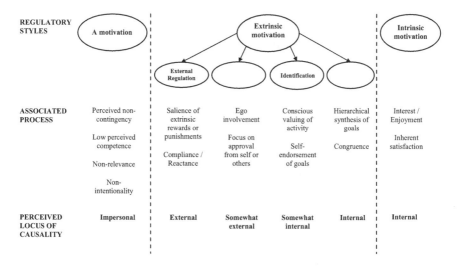

Figure 3.3   A taxonomy of human motivation
Source: Ryan, R.M. and Deci, E.L. (2000, p. 61)

Worker engagement is primarily linked to motivation to achieve specific goals. If the goals to be achieved are external then the probability that any individual will be 'engaged' in the achievement of these goals is directly related to the motivation they have to be aligned with these goals.

When individuals and teams are able to express their notions of risk in an organizational setting and those perceptions are seen to be given credence, it engenders affective states of belonging which is critical to worker engagement. Liang, Moreland and Argote (1995, p. 386) note that 'there is clear evidence that motivation depends not only on the objective characteristics of whatever tasks someone performs but also on that person's subjective evaluation of those tasks' and that 'these evaluations are often shaped by social influence processes within a work group'. It can therefore be hypothesized that worker engagement will be positively correlated to those factors that tend to motivate positively. In the context of the maritime industry some of these factors are thought to be satisfaction with income and nature of contractual relation with the employing company (job security). These social influence processes are further hypothesized to include the credence that is given by 'agents of the organization' to individuals' or teams' perceptions of the risk inherent to operations in the associated contexts, as well as individuals' or teams' ability to express these perceptions without inhibition.

Worker Engagement in Learning

The construct of worker engagement is used in this study to capture the degree to which seafarers are involved in decision-making regarding risk, how they are consulted (Cameron, Hare, Duff and Maloney, 2006; Nembhard and Edmondson, 2006; Shearn, 2004) and the commitment and effort they put into this themselves. This latter part involves 'the harnessing of organization members' selves to their work roles; in engagement, people employ or express themselves physically, cognitively, and emotionally during role performance' (Kahn, 1990, p. 694). Krause defines such engagement in relation to safety as individuals' connection to work intellectually, emotionally, creatively and psychologically (Krause, 2000, p. 24) and the UK Health and Safety Commission defines the term 'worker involvement' as:

> The development of relationships between workers and employers based on collaboration and trust and nurtured as part of the management of health and safety. This goes further than simply consulting workers. It involves a commitment to solving problems together. (Health and Safety Commission, 2006, p. 4)

The concepts as cited above is indicated diagrammatically in Figure 3.4

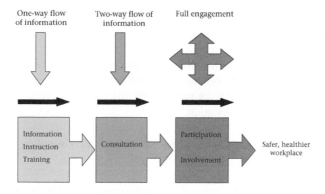

Figure 3.4   Worker engagement activities
Source: Health and Safety Commission (2006, p. 4)

The term 'worker engagement' is therefore used here to encompass the opportunities presented to workers to be involved in organizational safety goals, in the communication of risk perceptions and in processes of risk avoidance/ mitigation on the one hand, and the employees' commitment to do so on the other hand.

Figure 3.5 indicates (with minor adaptations) Cameron et al.'s attempt to show the traditional approaches used by organizations to engender worker engagement.

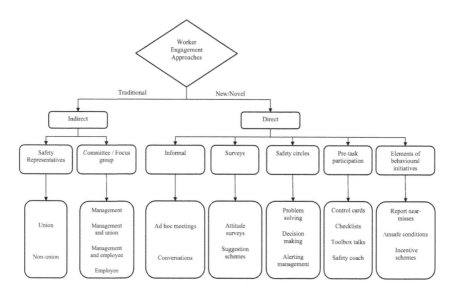

Figure 3.5   Approaches to worker engagement
Source: Adapted from Cameron, I., Hare, B., Duff, R. and Maloney, B. (2006, p. 7)

In summary of Cameron et al.'s (2006, pp. 7–9) explanation of the terminology in Figure 3.5:

- Safety representatives are often appointed or elected by different groups, for example, and may or may not be union affiliated.
- Safety committees may be made up of different permutations of manager and employee representation.

These two (as indirect approaches to worker engagement) are often the result of specific legislation:

- Ad hoc meetings are informal, non-obligatory contacts that serve to create forums for conversations between managers and operational staff.
- Surveys may take the form of paper-based media for example, questionnaires or suggestion opportunities. They are often without person-to-person contact, are anonymous and designed to be universal in outreach.
- Safety circles are purpose-specific setups that bring together volunteers to solve particular problems. They differ from safety committees in that they are short-term and do not have to meet at regular intervals.
- Pre-task participation requires the involvement of workers (who will be involved in a task) in a task-associated discussion before the carrying out of the task or mid-task when some change is intended. They involve a discussion of tasks within tasks and create opportunities for pointing out risk and for comparing existing risk assessment controls with the actual work in hand. Checklists or toolbox talks (where the discussion is centred round the equipment to be used) may be used to facilitate pre-task participation.
- Behavioural initiatives require certain specific behaviours by not only the workers but management as well. They often include goal/target-setting and feedback for near-miss reporting, days without lost-time-injuries (LTIs), speed of managerial response to employee concerns and meeting frequency. Their value is thought to be augmented by incentive schemes that go along with the feedback.

Shearn (2004, pp. 24, 25), in an analysis of the literature on worker participation (here taken to mean 'worker engagement') emphasizes the paucity of empiric data addressing many other factors relevant to worker participation in different industry settings.

> Worker participation espouses the values of democracy, autonomy, responsibility and trust, while it is often introduced within an employment context characterised by job insecurity, hierarchical power relations and short-term profitability. All this precludes the development of the values of [worker participation]. In this respect managers and employees may encounter and respond to [worker participation] initiatives that rarely meet with their expectations. As well as

seeking factors that might directly or indirectly contribute to the effectiveness of participatory processes, future research activities should focus on any tensions that can arise during implementation. (Shearn, 2004, pp. 27, 28)

This study aims, on the basis of these discussions, to examine such participatory and engagement processes specific to shipping organizations.

Chapter Summary

Continuing the discussion of the theoretical underpinnings of the research in this study which was started in Chapter 2, this chapter has focused on culture, organizational learning and the role played by motivation and worker engagement as discussed in the literature. It has sought to clarify some of the concepts that would be used as variables in the subsequent research. The chapter critically reviewed the literature on organizational learning and drew attention to some paradoxes of such learning as well as the role in learning played by how organizations give credence to different epistemologies of risk. The literature on organizational learning shows a merging of themes at the individual versus collective level, the kind of learning (experiential, vicarious, and so on), and the level of learning (single-loop, double-loop and so on). The literature also merges in broad themes of learning processes – knowledge acquisition, interpretation, distribution and organizational memory as well as learning products – strategy and action as a result of learning.

This chapter, like Chapter 2, notes divergences of views in the literature and indicates the theoretical position from which the research questions are derived. The subsequent research links the concepts – risk and its construal, organizational learning, team psychological safety and worker engagement – discussed in both chapters, a linkage that is not present in the literature. This it does in a context specific to the maritime industry, a context which is also not addressed by the literature.

Chapter 4
Research Questions, Methods and Measures

The significant problems we face cannot be solved with the same level of thinking we were at when we created them.

Einstein[1]

Purpose and Outline

The purpose of this chapter is to set out the global and specific research questions, the methods used for addressing them and the reasons for these choices. The theoretical concepts on which the research is based have been described in the two previous chapters. In this chapter the rationale for the operationalizing of these concepts are discussed. The approaches used in both the quantitative and qualitative phases of the study as well as issues of sampling and data use are presented. Limitations in the approach are identified together with measures taken to reduce the effects of these limitations. Also elaborated are some ethical issues deemed relevant to the study.

Research Focus and Questions

Chapter 2 discussed extensively the research on risk and its construal while Chapter 3 discussed organizational learning. Common to both these processes are social dynamics of psychological safety and engagement. However, there is not much research that links the concepts empirically and especially in the specific context of the maritime industry, which as has been pointed out, has unique characteristics even in the narrower context of high-risk industries. This study attempts to ameliorate this situation by addressing some of these empiric gaps.

Garvin, Edmondson and Gino (2008, p. 110) indicate that 'organizational research over the past two decades has revealed three broad factors that are essential for organizational learning and adaptability: *a supportive learning environment, concrete learning processes and practices, and leadership behavior that provides reinforcement* [emphasis added]'.

The discussions in Chapters 2 and 3 show the links in the literature that support this view as it relates to the construal of risk and organizational learning. Based on these theoretical foundations and considering the gaps in the literature as pointed out, the following research questions were derived from the global research goal, that is, finding out how shipping companies as organizations learn from, filter and

---

1   Retrieved 4 May 2009 from http://www.quotedb.com/quotes/11.

give credence/acceptability to differing risk perceptions and how this influences the work culture with special regard to group/team dynamics and individual motivation:

- What are the relationships between team psychological safety, leader inclusiveness, worker engagement and organizational learning as regards the context of shipping companies and ship officers?
- How do the constructs of team psychological safety, leader inclusiveness and worker engagement, predict organizational learning?
- Are the predictive potential of team psychological safety, leader inclusiveness and worker engagement for organizational learning confounded by variables such as rank, age, nationality and time in company?
- What variables influence worker engagement in ship officers as regards safety?
- What are the processes that facilitate or limit organizational learning in shipping companies?
- What factors influence credence-giving as regards risk information from ship officers?

In summary therefore answering these questions will help give insights into:

- How the relations between the constructs discussed affect the construal of risk, teamwork and possibly predict organizational learning in shipping companies.
- To what extent the existence of learning organizations (characterized by double-loop and deutero-learning) is manifested in the shipping industry.
- What the key considerations, processes and emergent issues in organizational credence-giving of risk perceptions are.
- How the giving of such credence and acceptability to different perceptions of risks influences organizational learning.
- To what extent there is an awareness of these issues and the need to optimize processes for greater risk resilience[2] in shipping.

Research Methodology: Mixed-Methods

The use of all forms of research methodology is limited by difficulties in achieving reliability, validity and meaning. In discussing this, de Vaus (2002, p. 55) notes

---

2   *Resilience* is the ability of a system to revert to a stable condition after being upset by disturbing factors with respect to risk, safety and reliability. Etymology – 1626, from Latin *resiliens*, 'to rebound, recoil,' from *re-* 'back' + *salire* 'to jump, leap'. Resilience. (n.d.). *Online Etymology Dictionary*. Retrieved 11 December 2008, from Dictionary.com website: http://dictionary.reference.com/browse/resilience.

that 'an awareness of the problem [of meaning and the difficulty of achieving reliability and validity] should encourage survey researchers to be more thorough in the way data are analysed and ... interpreted ... and to supplement their questionnaire studies with more in-depth data collection techniques'.

Mixed-methods research helps in complementing one method with another even where these methods are derived from different methodological positions. Taking entrenched and polarized methodological positions in research does not often do justice to the complexity of social analysis. Furthermore, the hard line disparity between qualitative and quantitative research is waning and there are increasingly more advocates for the more pragmatic and realistic research paradigm of mixed methods. Some view the 'war between the paradigms' as more a war of power than of knowledge. As Silverman puts it:

> The fact that simple quantitative measures are a feature of some good qualitative research shows that the whole 'qualitative/quantitative' dichotomy is open to question ... Many such dichotomies or polarities in social science [are] highly dangerous [being at best] pedagogic devices for students to obtain a first grip on a difficult field – they help us to learn the jargon. At worst, they are excuses for not thinking, which assembles groups of researchers into 'armed camps', unwilling to learn from one another. (Silverman, 2005, p. 8)

We neither have to be bound by Durkheimian positivism (see Giddens, 1972) nor by Weberian 'Verstehen' (see Runciman, 1978). Practicality demands that methods are blended and used to best ends for the research focus at hand (Martin, 2000, pp. 239–49). Furthermore, the interdisciplinary nature of most current social research – the present one not excluded – makes it almost imperative that a combination of the traditional mono-methods be used to limit their weaknesses and seek to merge their strengths. An increasing number of researchers tend to see quantitative and qualitative methodologies as extremities of a continuum (Johnson and Onwuegbuzie, 2004; Patton, 2002; Trochim and Donnelly, 2007). Extremities of a continuum are definitely important in many instances, but in many other cases, other points on the continuum will be relevant. Optimal pragmatic research – where it is intended that the findings be considered for practical value and application – most often lies somewhere on this continuum, merging appropriately aspects of the mono-methods and helping to avoid the situation where cumulative research entrenches partial views of phenomena as complete views. Accordingly a sizeable proportion of the research in the relevant fields being addressed here have resorted to mixed-methods or at least a multistage methodology (see for example Edmondson, 1999; Nembhard and Edmondson, 2006; Rossman and Wilson, 1985, 1991). It is said that Albert Einstein had a sign in his Princeton University office that read: 'Not everything that counts can be counted, and not everything that can be counted counts'. Mixed methods research seeks to get to what counts (including those that can or cannot be counted). The shipping industry is a very practical field and for academic research to have a positive effect, any research-to-practice gap must be addressed. This means that a method that is descriptive (qualitative) is required to enhance the dissemination and migration of findings from the insulated world of research to the practical world of shipping. In the latter world, processes are multiply determined by 'demands of external stakeholders and by the interests of internal actors, all of which must be addressed simultaneously' (Rosenheck, 2001, p. 1609) and research must appreciate this 'great conversation' of the larger world (Seale et al. as cited by Silverman, 2005). The sole reliance on quantitative data and conclusions based on them, necessarily controlled for many factors, may not be ideal for the real world of shipping.

In this study and in keeping with the theory of structuration, using both qualitative and quantitative steps also means that the unit of analysis could be shifted between individual ship officers and organizational practices as a whole.

The alternative use of the experimental research method was considered and discarded. This would have involved longitudinal studies of a number of teams, planned experimental interventions and the use of other teams for control purposes. Limitations in time, ethical issues, the effect on ship teams of such interventions and the practical difficulty in creating controlled conditions for a number of teams on board different ships in a longitudinal framework, made it impractical to use this method.

It is the case that the strongest case for causality between variables is made with experimental data. Without using this approach therefore, this study avoids stating categorical findings of causality, but limits itself to descriptions of and correlations between variables (questions of 'what' and 'how'). With the acknowledgement that 'correlational data cannot conclusively prove causality' (Hill and Lewicki, 2006, p. 3) any discussion of causality here is logical rather than statistical (Tabachnick and Fidell, 2007, p. 122).

The above constitutes the rationale for the employment of non-experimental mixed methods research (in a multi-step fashion) as the methodology for the study.

Exploring the Subject Area: Focus Group

The first step in the research effort was a focus group session with students of the World Maritime University (WMU) in Malmo, Sweden. The group was made up of 14 members from 10 different countries and with varying exposure to seafaring. All were enrolled in the WMU MSc. Maritime Affairs programme and work in the maritime industry. There were two females and 12 males; six participants had significant sea experience (management level ship officers) and three had none. The remaining five had some limited shipboard experience.

After a preliminary introduction to the subject area, views were solicited regarding the following questions:

- In your view what will show the existence of team psychological safety in a team on board ship?

- What operational areas on board ship can be affected by a lack of effective teamwork and team/organizational learning?
- The current organizational structure on board conventional ships is very hierarchical despite the very functional nature of shipboard work, as compared to civil society.
  - Is this a fair assessment of the status quo?
  - Would a flatter hierarchical structure be possible and advantageous?
  - What would be the impediments to a flatter hierarchical structure?
  - How would it be accomplished?
- What conditions and activities at the individual, team and organizational levels will, in your opinion, improve team psychological safety as a factor in organizational learning and risk construal?

There was general agreement in the group that all areas of shipboard work are affected by teamwork and hence by team psychological safety in relation to learning.

It was pointed out that due to the unique nature and environment of shipboard work, teams have to exhibit positive team dynamics despite the fact that their periods of activity do not always coincide, for example in the watch system. Teamwork was understood not to be relevant only for co-located and co-temporal teams but for all entities working for common goals in the context of defined structures across time and space.

At a higher level, therefore, effective teamwork and team psychological safety was just as relevant to the ship–shore interface as it was to the micro-level teams on board ship.

The group felt that the nature of meetings, the structure of teams, the degree to which a blame culture existed, the existence of feedback loops, communication patterns, individual perceptions of team empathy, nature of work, task roles and delegation, retention/attrition rates and so on could all help as indicators of the state of team psychological safety and optimal teamwork. The relevance of leader inclusiveness was pointed out, as was the place of individual assertiveness and importance of managerial commitment.

The group was of the opinion that crew engagement through organizational policies and education would help organizational learning. Some in the group felt that the nature of work on ships was such that the extant rigid hierarchy was the best suited structure. Others thought it was perpetuated by tradition and that other formats could work just as well. A number of discussants pointed out what they felt was the role that culture has to play in assertiveness and vocal contribution to the team. While all agreed that speaking up was important, the reasons for some team members not doing so, even when they perceive grave risk, were considered to be varied. Some of the reasons mentioned were perceptions of the team member about their own inexperience as it relates to rank, their nationality, consequences of earlier attempts to speak-up (the history of organizational/superior credence-

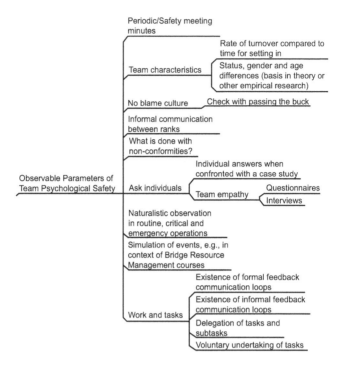

Figure 4.1   Focus group notion of observable outcomes of TPS

Note: The figure was generated using MindManager X5, a 'brainstorming' software used to facilitate the focus group session.

giving to their perceptions) and the value they feel is placed on their contribution to the team by other team members (and the larger organization).

Regarding possible observable indicators of team psychological safety and ways of assessing these, Figure 4.1 shows some of the issues raised by the focus group members.

An interesting observation during the focus group exercise was the play out of part of the phenomenon being discussed in the setting of this group. It was almost like a planned role play, where certain individuals assumed more vocal and dominant roles than others. The forming dynamics of this particular group had led to the acknowledgment (albeit tacitly and perhaps subconsciously) of certain individuals as 'leaders' (based on such variables as sea-experience, current work, language ability and personality). This meant that there had to be some control on the part of the facilitator to get others heard and even then some did not speak up. It is interesting to note that two of these approached the facilitator after the session to discuss their opinions and also to try to explain their silence.

This focus group session served to flesh out some of the issues raised in the literature and to clarify further the desired approach to the global research question.

Quantitative Survey: Participants, Procedures and Measures

The unit of analysis in the quantitative phase of the research was the ship officer and his/her perceptions. A structured questionnaire intended to measure four constructs derived from the theory/literature discussed in Chapters 2 and 3 and the focus group outcomes, was designed. These constructs were chosen based on their role in organizational behaviour in risk construal and learning. They were:

1. Team psychological safety (TPS) 'The shared belief held by members of a team that the team is safe for interpersonal risk-taking' (Edmondson, 1999, p. 350) and where these members feel 'able to show and employ [themselves] without fear of negative consequences to self-image, status, or career' (Kahn, 1990, p. 708).

2. Leader inclusiveness (LI) 'Words and deeds by a leader or leaders that indicate an invitation and appreciation for others' contributions. Leader inclusiveness captures attempts by leaders to include others in discussions and decisions in which their voices and perspectives might otherwise be absent' (Nembhard and Edmondson, 2006, p. 947).

3. Worker engagement (WE) 'The harnessing of organization members' selves to their work roles ... [where] people employ or express themselves physically, cognitively, and emotionally during role performance' (Kahn, 1990, p. 694) ... [Worker engagement] is used to encompass the opportunities presented to workers to be involved in organizational safety goals, communication of risk perceptions and processes of risk avoidance/mitigation on the one hand, and the employees' commitment to do so on the other hand.

4. Organizational learning (OL) The intentional or unintentional processes by which organizational social systems (made up of individuals and groups) create the potential (in cognitive, behavioural and/or affective terms) to adapt in behaviour, values and attitudes as a result of exposure to experiential, vicarious or contextual events whether or not there is a conscious awareness of this process. Learning also includes the process of hypothetically/rationally inferring causal links not explicitly based on historicity but linked to patterns stored in memory.

'Organizational learning occurs through shared insights, knowledge, and mental models ... Learning builds on past knowledge and experience – that is, on memory' (Stata, 1989, p. 64). 'Organizations are seen as learning by encoding inferences from history into routines that guide behaviour' (Levitt and March, 1988, p. 319). It can also be seen as 'the way firms build, supplement and organize knowledge and routines around their activities and within their cultures, and adapt and develop organizational efficiency by improving the use of broad skills in their workforces' (Dodgson, 1993, p. 377).

All reference to organizational learning (OL) in the quantitative phase of the research is a reference to organizational learning as perceived by the respondents.

In addition to these four constructs, data relating to the respondent's perception of organizational credence-giving to risk perceptions of individual/team and organizational learning was sought, together with a number of other variables. Opportunities for comments were also given via open questions in the survey/ questionnaire.

Questionnaire items for the four main constructs were drawn from validated studies by earlier researchers in the different domains. Items used in the assessment of the constructs of team psychological safety (TPS) and leader inclusiveness (LI) were based on the work of Edmondson and Nembhard (Edmondson, 1999; Nembhard and Edmondson, 2006). The LI items used were selected in preference to Graen and Uhl-Bien's scale items for leader-member-exchange (Graen and Uhl-Bien, 1995) since these did not exactly capture the construct being examined in this study. Research work by Dorai, McMurray and Pace (2002) was the basis for the selection of items for the construct of organizational learning (OL). For worker engagement (WE) the work of Cameron et al. (2006) and the Gallup Workplace Audit as cited by Harter, Schmidt and Hayes (2002), was found relevant. Cameron et al.'s research effort also gave the theoretical basis to associate this construct with that of 'perceived organizational support' (Eisenberger, Huntington, Hutchison and Sowa, 1986). The initial items were therefore generated and adapted from the specific scales used by the different researchers and as indicated in the literature. Adaptations were informed by the particular context of the maritime industry and results of feedback from the focus group as well as an expert group and pilot run which are described next.

Each construct was assigned a number of scale items which comprised statements to which respondents had to note their degree of agreement in a 5-point Likert response format as follows:

1 – Strongly disagree; 2 – Disagree; 3 – Neutral; 4 – Agree; 5 – Strongly Agree.

The full Likert Scaling process (Singleton Jr. and Straits, 2005, pp. 273, 390–91; Trochim and Donnelly, 2007, pp. 102, 136–7) was not required in generating scale items because the items were based on the literature and pre-existing validated scales and adapted only for the maritime industry. Instead, the initial questionnaire was then sent out to a group of experts (7 subject area industry experts and critical minds from academia) for independent and unbiased reviews. The use of this group was to improve the validity of the survey methodology (Trochim and Donnelly, 2007, p. 50) and to gain a critical evaluation of the content and methodology of this particular part of the research approach (Barnett, Stevenson and Lang, 2005). Correspondence with the group was electronic. Group members were unknown to each other as group members; neither did they have any study-relevant contact with each other.

Subsequent to the receipt of insightful critiques and feedback comments from this panel, the questionnaire was amended and then sent out on a pilot test (de Vaus, 2002; Singleton Jr. and Straits, 2005). Fifty-two responses were received, reflecting to a degree the final desired stratified random sample of the target population based on the seafarer demographics in the latest update of the BIMCO/

ISF study on the worldwide demand/supply of seafarers (Warwick Institute for Employment Research, 2005).

The information sought from this pilot survey included a determination of whether:

- All the questions were commonly understood.
- All the questions were interpreted similarly by all respondents.
- All the questions had an answer that could be marked by every respondent.
- Each respondent was likely to read and answer each question.
- Respondents had any comments about the questionnaire.

It was possible to get this feedback because the questionnaires were administered in the presence of this researcher or others with an understanding of the issues being addressed.

The results of the pilot run were statistically analysed using the computer software Statistical Package for the Social Sciences (SPSS) version 16. An analysis was made for reliability using inter-item correlation (Cronbach's alpha) for each construct. The resulting Cronbach's alphas for the constructs in the pilot were:

- TPS (12 items)   = 0.476
- LI (10 items)   = 0.749
- OL (12 items)   = 0.844
- WE (13 items)   = 0.620.

Social science and organizational research literature recommend Cronbach's alpha values to be greater than 0.7 for inter-item correlation (de Vaus, 2002; DeVellis, 2003; Dewberry, 2004, p. 321). As per this criterion the results were checked for items which when deleted would increase the reliability of the scales. For TPS and WE, deletion of items only marginally increased the Cronbach alphas. Accordingly, and based on queries about ambiguity and response bias in questions from the expert group, a few of the items were eliminated and others refined. Five items were reverse-coded to limit response bias (Pallant, 2007, p. 83), specifically position bias (Singleton Jr. and Straits, 2005, p. 294) because of the response format used. These resulted in achieving final Cronbach's alphas of 0.75 for TPS, 0.82 for LI, 0.90 for OL and 0.88 for WE when the final questionnaire was administered to the research sample.[3]

The revised questionnaire included nine items for TPS, nine for LI, 11 for OL and 12 for WE with all items rated on a five-point Likert response format ranging from 'strongly agree' to 'strongly disagree'. The questionnaire also had one item regarding the perceptions of organizational credence-giving to risk information based on a number of factors. There were additionally 19 items[4]

---

3   See Table 5.1 in the next chapter (Chapter 5: Research Findings).
4   See Appendix 3 for descriptive statistics of these variables.

related to respondent demographics including age, nationality, gender, rank, time spent in present company, satisfaction with salary, mode of employment, nature of employment contract, time spent at sea, time spent in shore management, experience as a rating, experience with a multinational crew, size of company fleet, country of training, flag of last ship, enjoyment of working at sea, enjoyment of working with company, last ship, most time ship. Finally a number of opportunities for open respondent comments were given.

This questionnaire was sent back to specific experts in the original expert group for review, in the light of their earlier comments. At this stage no further comments were obtained. The final version can be seen in Appendix 2.

Final administration of the questionnaire to target research participants was then done via email, distribution of hardcopies and by an internet-based survey site.[5] The use of a multi-mode means of survey administration was considered beneficial as indicated by Dillman (2007) and the pilot study suggested that no substantive variation in responses existed as a result of the different administration modes.

An attempt was made to get data sources that would give a sample reflecting the demographics of the global seafarer population especially regarding nationality. The method used in this step was stratified random sampling (de Vaus, 2002; Singleton Jr. and Straits, 2005) using the percentages of the seafarer population as indicated in the BIMCO/ISF study of 2005 (Warwick Institute for Employment Research, 2005) and as shown in Figure 4.2. This was done by initially receiving results in a totally random manner and then targeting specific geographical regions (albeit surveying in a random manner) in an attempt to replicate as closely as practicable, the percentages indicated in the BIMCO/ISF study.

Qualitative Studies: Field interviews and System Analysis

With the rationale for the choice of methodology in mind the initial studies were augmented[6] with field studies in a number of shipping company offices for sessions of in-depth interviewing and software/document analyses. The unit of analysis in this phase of the research were the organizations themselves – an examination of their recurring practices in time and space.

---

5   The site was hosted online and could be accessed via the link: http://portals.wmu. se/intranet/ResearchConsultancy/PHD/SurveyMaritimeOrganizationalLearning/tabid/247/ Default.aspx. The survey site required login. Username for login was: seasurvey and password: seasurvey08.

6   With respect to the global research question and not as an addition to the multi-mode administration of the survey questionnaire.

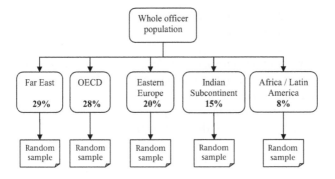

Figure 4.2   Diagrammatic representation of stratified random sampling (percentages) of global officer population derived from BIMCO/ISF data

Theoretical Orientations

As indicated by Rosness et al. 'it is important to study how organizations handle their daily operations, correct deviations and learn from normal and abnormal situations' (Rosness, Guttormsen, Steiro, Tinmannsvik and Herrera, 2004, p. 8). In this context, the conceptualization of organizational learning included the process of learning, the context of learning and the outcome of learning (Naot, Lipshitz and Popper, 2004, p. 452). The structured and intentional use of organizational learning in the context of shipping companies was examined with reference to the following: existence, breadth, elaborateness and thoroughness (Huber, 1991, p. 89) as well as the framework for the social 'amplenuation' of risk (Kasperson, Kasperson, Pidgeon and Slovic, 2003).

Following Templeton et al. (2004), three theoretical themes and two broad operational themes are used to subsume five organizational learning constructs. The *three theoretical themes* are:

1. Individual: organization (collective) interaction (levels of societal aggregation).
2. Levels and types of learning: for example, single-loop learning, double-loop learning and deutero-learning (levels) and congenital, experiential, vicarious, contextual, and inferential learning (types).
3. The role of knowledge structures.

The two operational themes relate to:

1. The domain of learning **processes** that cover Huber's four constructs (Huber, 1991) as indicated in figure 3.2:
   – Knowledge acquisition

-   Information interpretation
-   Information distribution
-   Organizational memory.

2. The domain of learning that includes **adaptation and structural change** (as a result/outcome of learning) resulting in a fifth construct:
    -   Structural change/adaptation.

The themes and their sub-themes clearly map unto the Social Amplenuation of Risk Framework. As such the SARF is theoretically incorporated in the schemes.

Methodological Approach

The aim of this phase was to qualitatively seek meaning and deeper nuances regarding relevant organizational behaviours, beliefs and attitudes related to these themes and constructs of organizational learning and risk construal, the determination of which may be limited from survey outcomes focusing on ship officers. The field was approached with a loose design regarding the conceptual framework of the constructs in mind and in reference to the theoretical and operational themes of organizational learning and to the social amplenuation of risk framework. As suggested by Miles and Huberman (1994, p. 17 especially paragraph five), it was not considered optimal to approach the field with a blank slate (completely loose design) in the quest for the generation of purely inductive research questions. The conceptual framework was derived from the literature and is essentially deductive (Berg, 2001, p. 6; Lewins and Silver, 2007; Miles and Huberman, 1994). However, a degree of looseness was retained in order to introduce an inductive element into the research and to allow for the unearthing of emergent themes in the study area. The resulting tighter design brought clarity and focus.

Using a purposive sampling approach, a number of organizations were selected. Visits of up to four working days were paid to these organizations and in-depth interviews done with key safety and operational staff together with document and software perusal.

The shipping companies[7] were selected based on the following:

-   Espoused change and learning based on the occurrence, over a short period, of significant accidents in otherwise decades of safe operations.
-   The perceived role of certain dynamics of accident causation relevant to the current study such as team psychological safety.

---

7   The shipping companies and interviewees are all made anonymous by the use of fictitious names. In some instances non-relevant details not affecting the substance of the research outcome have also been altered. Any similarity between these names/details and those of any other existing companies who were not part of this research, their employees and/or others, is not intended, and does not in any way imply a reference to these companies, employees and/or others.

- global ship management role.

As Cox and Flin (1998, p. 194) note, 'organizations who have suffered major accidents … organizations who operate in a hazardous environment but apparently have low accident rates (high-reliability organizations) and organizations who are experiencing extensive change', are ideal for qualitative studies into safety, using – among others, in-depth interviewing and observation. It was deemed that this informed opinion was relevant to the current study and that the organizations selected reflected, at one point or another in their existence, the characteristics indicated by Cox and Flin.

Description of companies   Olive Shipping is a Norway-based shipping company. The company was established in the early twentieth century and at the time of this study operates and manages just less than 100 ships with direct ownership of about two-thirds of these. The ships operate mainly in the specialized tanker trades and are managed from two offices in Europe and Asia. The crew, since the mid-80s, have also been drawn from these two regions. The company has established its own training centre. Olive Shipping was chosen due to the recent occurrence of a major accident on an owned vessel, which accident was characterized by dynamics reflecting the constructs under examination. The accident occurred after decades of relatively high standards in safety. Interviews were conducted at the site in Norway with three senior level personnel – Paul (quality and risk management advisor), Andrea (crew welfare officer) and Barry (leader/manager of all marine superintendents in the fleet). Additionally a telephone interview was conducted with Edith (crew affairs and legal advisor) in Asia. There was significant document and software perusal in the context of the interviews with Paul and Barry.

Similar to Olive Shipping, the Magna Group has at the time of the study, enjoyed relatively long periods of high-level safety in ship operation. This Asia-originated group has many offices in different global locations. The research visit was to the London offices of three shipping companies in the group – Alpha Tankers Ltd, BetaGas Carriers and Gamma Shipping. Though the three belong to the larger group and thereby share (in theory at least) the same group philosophy regarding safety and risk – as well as certain physical emergency resources – the companies operate as distinct, separate and independent entities. This group as a whole was chosen due to the occurrence of a series of accidents in a particular year, despite many earlier years of relatively high operational safety levels. Like Olive Shipping, and as a result of the perceived significance of these accidents, the organization reported the undertaking of a major review of its operations across all divisions, with significant implications for how risk was viewed and managed. Studying companies in a group also helped to investigate organizational culture effects at more organizational levels. The first interviews were conducted at the offices of Alpha Tankers Ltd, a company that operates and manages 28 tankers. The interviews were done on site with Jason and Tim. Jason is the deputy manager and Designated Person Ashore (DPA) and Tim is a superintendent in charge of health, safety, environmental and quality issues. During the interview with Tim, a significant number of documents were made available and the software in use for the area of examination was extensively reviewed.

In the offices of BetaGas Carriers, an interview was done with Evan, the managing director. Evan has 18-years of experience in senior management on ships and also significant experience in operations and management in the Liquefied Natural Gas (LNG) sector. The interview was done while together examining some of the documents and software the company uses. BetaGas Carriers currently manages and operates seven specialized tankers with three more on order.

An interview similar to the ones undertaken in Alpha Tankers and BetaGas Carriers was done with Akemi, a manager in Gamma Lines. This company specializes in the carriage of neo-bulk cargo and manages about 100 ships.

The next organization, Chrestos LNG was chosen because this company was an offshoot of an earlier management group that was in charge (as managers) of the particular ship involved in the major incident mentioned earlier in relation to Olive Shipping. Chrestos LNG in its current form was incorporated in 2005 and operates and manages about 10 vessels in a world wide trade with four on order. An interview was done on site with Alexander, the manager in charge of health, safety, security and environmental issues and also Designated Person Ashore (DPA). In the setting of the interview, some documents were reviewed.

The last organization visited was Metron Shipping. This company reported no major accidents in their time of existence in the current form. Metron, established in the very early 1970s, currently manages (including worldwide operation, maintenance and crewing) an overall tonnage of about 4 million deadweight (DWT) comprising of about 30 tankers (VLCCs, Suesmax, Aframax tankers and product carriers) and 10 bulk carriers. The company has a number of new buildings on order. An interview was granted by Vasilis, the head of the quality and safety department.

All the companies studied were certified as ISM compliant and additionally had certification for ISO 9001.

A brief summary description of the companies and interviewees is shown in Table 4.1.

Access to Olive Shipping was obtained by placing a direct call to the relevant department in the organization. For the rest of the companies, access was obtained through introductions by third parties.

In all nine face-to-face and one telephone in-depth interviews were conducted in 2008 and 2009 at six sites. Together with the interviews, significant document and software perusal was undertaken. The time spent at any particular site ranged from a day to four days. The interviewees were mainly drawn from general management, predominantly in departments overseeing health and safety, risk, environmental issues, quality and crew welfare. The interviews themselves, lasting on average one hour each, were semi-structured and designed to lack the rigidity of formal interview sessions. They were essentially conversational, though guided by the overall theoretical framework, and allowed the respondent to structure their

Table 4.1   Outline of organizations and interviewees

Company	Description	Interviewees	Role and background
Olive Shipping (Oslo)	Norway-based Established early twentieth century. Operates and manages about 100 ships	Paul	Quality and risk management advisor – No seafaring background – Degree in Maritime Shipping Business.
		Andrea	Social and Welfare Officer – Background in health industry – No seafaring background.
		Edith	Head of Crew and Family Affairs – Seafaring experience on cruise ships as ISM and Safety Coordinator.
		Barry	Manager – Head of Superintendents – Seafaring experience as deck officer.
Magna Group	Asian origins – 1917. Among top 10 carriers of the world		
Alpha Tankers (London)	Manage 28 tankers	Jason	Deputy Manager and DPA – Seafaring experience as engineer officer.
		Tim	Superintendent in charge of HSEQ – Seafaring experience as deck officer.
BetaGas Carriers (London)	Manage and operate seven LNG tankers	Evan	Managing Director – Seafaring experience as master.
Gamma Lines (London)	Neo-bulk shipping (car carriers). Operate about 100 ships	Akemi	Deputy General Manager – Seafaring experience as master.
Chrestos LNG (Piraeus)	Established March 2005. Manage 10 LNG vessels	Alexander	Manager in charge of HSSE, DPA – Seafaring experience as master.
Metron Shipping (Piraeus)	Established early 1970s. Manage 40 ships	Vasilis	Manager, Safety and Quality Support department – Seafaring experience as master.

own accounts. As suggested by Miles and Huberman (1994, p. 50) forays into data analysis commenced even before all the interviews were done so that the possibility of looking at issues in new ways was retained in latter interviews. To gain the necessary insights, the analyses relied on the interpretations of the people interviewed with due cognizance of their roles in the target organizations.

Focusing on six shipping companies may be seen as too restrictive and raise questions about the external validity of the findings of this research stage as well as about reliability. The study chose to use this approach based on the merits of being idiographic in an attempt to raise the relevant issues by in-depth study of a few cases (Simon, 1991, p. 133) – a collective case study 'where a number of cases are studied in order to investigate some general phenomenon' (Silverman, 2005, p. 127).

The choice of these organizations does not imply that they are culturally characterized by deviance or evasiveness. On the contrary, these are companies recognized for adherence to international regulatory standards. They have better than average safety-systems and for the greater part of their existence fall into the category of organizations which, despite their operation in contexts that are arguably high-risk, apparently have low accident rates. They were chosen on the rationale that they epitomize the higher end of compliance (but not verifiably the highest end of an excellence culture). The focus, like that of Marcus and Nichols (1999, p. 485) for the nuclear industry, is on 'organizations that typically, but not always, operate within a safety border' and with rare major casualties. Measures that work in this band of shipping are more likely to work across the board. This band is also important because one publicized accident occurring in this band is likely to increase recalcitrance, organizational recreancy, deviance and evasiveness in lower-end operators. The latter kind of operator simply draws the conclusion that a safety/quality focus is not worth the effort in time and other resources. Accordingly while the purposive selection of some of these companies is premised on the occurrence of significant accidents in otherwise 'safe companies', this selection is not meant to suggest that the companies are accident prone: quite the contrary is implied.

After the interviews were conducted, significant time was spent transcribing, cleaning up and making the interview data anonymous. Paid-for transcription services were not used. Among other reasons and importantly, ethical research issues regarding anonymity did not allow for the divulging of all the statements in the 'raw' form to a transcription service. Fortunately, the transcription process was beneficial to the writer in the sense that it increased early exploration and analysis of textual data. As noted by Lewins and Silver (2007, p. 57), 'reflection during and after data gathering and while transcribing forms part of the overall process of exploration, and for many this is an important phase of analysis'.

The transcribed data from the interviews as well as from a number of documents were analysed with the aid of the Qualitative Data Analysis (QDA) software, Atlas.ti version 6.0.

Additionally answers to the open-ended questions in the quantitative analysis phase (from the earlier questionnaire/survey and essentially qualitative) were analysed in the QDA.

Based on the theoretical approach as discussed with respect to organizational learning and SARF (see Figure 2.1 and Figure 3.2), analysis of the data was initially approached with the assigning of a predetermined set of codes and sub-codes derived from the literature (see for example Holland, Holyoak, Nisbett and Thagard, 1989; Huber, 1991; Kasperson et al., 2003; Templeton et al., 2004; Tyre and von Hippel, 1997) and as shown in Table 4.2.

Table 4.2   Predetermined coding for theoretical/operational themes

Code	Sub-code: level 1	Sub-code: level 2
Knowledge acquisition (sources of information)	Congenital learning	
	Experiential learning	Organizational experimenting
		Organizational self-appraisal
		Unintentional learning
		Unsystematic learning
		Experience-based learning curves
	Vicarious learning	
	Grafting	
	Inferential/inductive learning	
	Contextual learning	
	Searching and noticing	Scanning
		Focused search
		Performance monitoring
Information interpretation	Cognitive maps and framing	
	Media richness	
	Information overload	
	Unlearning	
Information distribution	Information channels	
	Social stations	
	Individual stations	
Organizational memory	Storing of information	
	Retrieval of information	
	Computer-based organizational memory	
	Transactive memory	

Table 4.2   *Continued*

	Single-loop learning
Level of learning	Double-loop learning
	Deutero-learning
Change and adaptation	

After the initial exposure to the data gained from the interview interactions as well as the transcription process, a more rigorous and reflexive two-stage coding run was done on each interview. Codes were assigned as per their relation to the theoretical themes. Emergent themes were newly coded and links based on theory and empiric findings created, followed and further analysed. All codes used in the qualitative data analysis (QDA) were primarily used as markers for relevant quotes. The quotes themselves are pointers to underlying structures of the theoretical and emergent concepts and were not necessarily definitive in relation to the associated code.

Research Ethics

As per acceptable research standards, the interviews were done and voice-recorded with the informed consent of all interviewees. Prior to all interviews, the nature of the study, the particular purpose of the interviews and the use to which data was to be put were made clear to the participants. All participants were willing to grant the interviews under the circumstances and confirmed their informed consent with signatures on the relevant form. In accordance with the agreed requirements, all companies and names of interviewees have been made anonymous in this text.

Possible Limitations of the Study

- The focus of this work has been necessarily limited to specific parts of what is, holistically speaking, a profoundly complex and unstructured issue (Dörner, 1996; Maier, Prange and Rosenstiel, 2001, p. 21). The study acknowledges this complexity and its own limitations in the sense that it is unable to cover in greater detail the wide spectrum of all the factors that contribute to organizational learning, risk amplenuation in social settings and safety management. These factors include economics (macro and micro), individual behavioural tendencies (in the realm of micro-level psychology), the politics associated with social groupings, national legal and policy frameworks as it affects the environment within which shipping organizations operate and the supply and demand dynamics of shipping which affect human resource retention and integration. Organizational learning factors as discussed in this book are not the only determinants for organizational safety performance and risk management.

- In the survey, items for organizational learning do not in reality measure the organizational learning of the respondent's company per se, but actually the respondent's perception of organizational learning in the company. This limitation is acknowledged and it is again stressed that the study considers the perceptions as equally important, noting the issues raised in the literature regarding the subjectivity of risk.

- The mixing of the main research paradigms may be seen by some as compromising the strengths of each, a position not shared by this author as discussed earlier.

- A limitation of the qualitative phase of the study – as with a lot of qualitative research – is that the interviews may be seen as portraying the subjective views of the interviewees. It is acknowledged that biases may be characteristic of subjective reports and opinions. Magala (2004, p. 13) notes the difficulty researchers face – when using attitude surveys in the investigation of values, beliefs and attitudes – in differentiating between 'reality' and 'desirability': the difference between 'values actually referred to and operationalized by respondents into norms or counter-norms in real life situation versus values they see fit to declare, but not necessarily to follow, especially in some specific circumstances'. This has been called 'espoused theory' and 'theory-in-use' by Argyris and Schön (1996, p. 13). The effect of this is limited by the inclusion, where possible, of evidence in the form of documents and software into the analyses to increase objectivity. These serve as records of recurring practices. The positions of the individuals interviewed, especially for those of them who are influential management actors, also brings further validity to their statements as a reflection of the views of the particular organization. Additionally what was particularly relevant are the perceptions of the research subjects since it is these perceptions that inform the recurrent practices key to the process of structuration.

- It has been noted that 'real-world decision-making processes are rarely well documented, and it is hard, if not impossible, to reconstruct them. Reports of real life processes of this kind are often unintentionally distorted or even intentionally falsified' (Dörner, 1996, p. 9). A fundamental premise of the study was that interviewee and questionnaire respondents answered all questions and made statements that were, to the best of their knowledge, an accurate presentation of their perceptions and that they were not subject to any institutional pressure to respond in particular ways. Scales such as the Marlowe-Crowne Social Desirability Scale (Crowne and Marlowe, 1960; Strahan and Gerbasi, 1972) were not used to avoid making the questionnaire inappropriately long and thereby reducing response rate.

- Access to the target population via the use of the internet-based questionnaire was limited due to the recognized current low levels of internet services on ships. Other modes of questionnaire administration were used in an attempt to reach as many ship officers as possible.

- The background of the author – a seafarer (master mariner) – may be suggestive of a bias during the research. Efforts have been made to limit any such bias. These include the setting out of a rigorous and transparent research methodology and process. On the positive side, this background also meant that the researcher could bring an understanding to the context based on a level of syntactic and semantic knowledge which perhaps will not be readily available to organizational researchers not familiar with the shipping industry.

These limitations have been borne in mind throughout. The research method and measures taken to ameliorate them have included careful setting out of research limits, scrutiny of data, validity and reliability tests as well as multiple sources of data and mixed methods as a research methodology.

Chapter Summary

The focus of the chapter has been on the research questions, the methodological approach and the measures taken. A number of specific research questions were presented as expanding on the basic research inquiry into how shipping companies as organizations learn from, filter and give credence/acceptability to differing risk perceptions and how this influences the work culture with special regard to group/ team dynamics and individual motivation.

The mixed-methods (multi-step) approach was described and the rationale for its use given. The process included a focus group discussion, generation of survey items for a questionnaire covering the relevant constructs, the validation of the items using an external reference group of experts, and a pilot run. The questionnaire was then administered via hard copies, email attachments and online hosting. A stratified random sample based on ship officer nationality percentages from the BIMCO/ISF survey was targeted.

Subsequent to this step – and in a qualitative phase – a number of field studies were undertaken. They consisted of visits to six purposively-sampled shipping companies and the conducting of in-depth semi-structured interviews and software/document perusal. The interview structure was drawn from three theoretical themes (in relation to risk, its construal and amplenuation and learning) and two operational themes.

Finally the chapter addressed some potential limitations of the methodology, sampling and research as such.

Chapter 5
Research Findings

Most discussions of decision-making assume that only senior executives make decisions or that only senior executives' decisions matter. This is a dangerous mistake.

Peter Drucker[1]

Purpose and Outline

In this chapter the results of the research are indicated. The global research inquiry was into how shipping companies as organizations learn from, filter and give credence/acceptability to differing risk perceptions and how this influences the work culture with special regard to group/team dynamics and individual motivation.

The specific questions were:

- What are the relationships between team psychological safety, leader inclusiveness, worker engagement and organizational learning as regards the context of shipping companies and ship officers?
- How do the constructs of team psychological safety, leader inclusiveness and worker engagement, predict organizational learning?
- Is the predictive potential of team psychological safety, leader inclusiveness and worker engagement for organizational learning confounded in anyway by variables such as rank, age, nationality and time in company?
- What variables influence worker engagement in ship officers as regards safety?
- What are the processes and factors that facilitate or limit organizational learning in shipping companies?
- What factors influence credence-giving as regards risk information from ship officers?

Results from quantitative data analyses are presented first and involve descriptive and inferential statistics. This chapter also presents the results of data from the qualitative research approach in terms of the five operational themes: knowledge acquisition, information interpretation, information distribution, organizational memory and structural change/adaptation and the emergent themes. The results are presented in network diagrams showing coding densities and groundedness.

---

1   From the article 'What makes an effective executive,' in *Harvard Business Review*, vol. 82, no. 6, June 2004.

Quantitative Research Findings and Analyses[2]

Considering all survey administration modes, total received and answered questionnaires[3] numbered 132. Of these, four hardcopy responses were ineligible because the respondents were not from the target population. One hardcopy was deleted because it had more than 85 per cent of the variable data missing. Five responses (four from the internet-based mode and one hardcopy) were deleted because they were duplicated. This gave a final data set for analysis of N=122.

Apart from the one case with 85 per cent of data missing (which was deleted outright) a few random and non-significant missing values (ranging from 0.8 per cent to 4.8 per cent) found in the screened set of TPS, LI, OL and WE scales items were left in the data set to be handled as per the analyses they are used in (Tabachnick and Fidell, 2007, p. 63). The effect of five outliers on the total scores of these scales was moderated by changing the relevant values by one each. Twenty-eight values out of the total of 5002 (0.6 per cent) were so moderated.[4]

Though the final percentages for regional representation obtained (see Figure 5.1) did not track exactly with the BIMCO/ISF percentages they were considered sufficiently representative of the global ship officer population especially in light of the Drewry Manning Report for 2008 (Precious Associates Limited and D.M. Jupe Consulting, 2008, p. 19). This latter report puts the percentages at 20 per cent for Eastern Europe, 16 per cent for Western Europe, 36 per cent for the Far Eastern Bloc and 28 per cent for the rest of the world.

It is also noted that in social survey research, many reported percentages are a property of the particular survey at a moment in time.

The sample set of these final 122 respondents had a mean age of 37.0 years. There were:

- 118 males (96.7 per cent) and 4 females (3.3 per cent)
- 58 (47.5 per cent) operational level officers and 62 (50.8 per cent) management level officers and 2 missing values (1.6 per cent)
- 42 (34.4 per cent) less than 30-years-old, 41 (33.6 per cent) between 31 and 43-years-old and 39 (32 per cent) more than 44-years-old[5]

---

2   The analysis and reporting format are informed by Pallant (2007), Tabachnik and Fidell (2007) and the Publication Manual of the American Psychological Association [APA] (2001).

3   87 delivered hardcopies, six by email attachment and 39 responses to web-hosted survey. Given the nature of questionnaire administration, it was impracticable for a response rate to be obtained for the whole data set. For the hard copies sent out, the response rate was 82.1 per cent, calculated using the formula: Response Rate = (Number returned *100)/ [No. of hardcopies – (ineligible + unreachable)] i.e., (87*100) / (110–4) Ineligible refers to 4 responses that were sent in by respondents not in the target population.

4    $(N)_*(\text{Number of scale items}) = (122)_*(41) = 5002$.

5   The sample was divided into these age categories not based on fixed age intervals, but on equivalence of sample sizes.

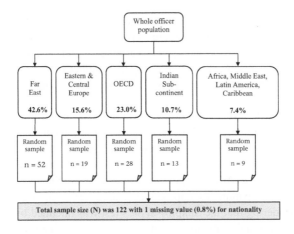

Figure 5.1   Stratified random sampling for research population

- 91 (74.6 per cent) working in the Deck department and 27 (22.1 per cent) in the Engine department and 4 missing values (3.3 per cent).

Detailed sample demographics and descriptive statistics for age, gender, nationality, rank, and other variables are indicated in Appendix 3.

An analysis for reliability for the final sets of items for each theoretical construct yielded Cronbach alpha values as indicated in Table 5.1.

Table 5.1   Scale reliability (Cronbach alpha) for measured constructs

	Cronbach's alpha	Cronbach's alpha based on standardized items	N of items	Similar scales[a] in literature
TPS	.753	.775	9	0.73[b]; 0.82[c], 0.71[d]
LI	.819	.827	9	0.75[e], 0.80[c]
OL	.902	.905	11	0.95[f]
WE	.877	.885	12	0.91[g], 0.97[h]

[a] The comparison scales are not exactly the same as those used in this research but are adapted to various degrees for this context.

[b] Nembhard and Edmondson (2006).

[c] Edmondson (1999).

[d] May, Gilson and Harter (2004).

[e] Nembhard and Edmondson (2006).

[f] Dorai, McMurray and Pace (2002).

[g] Gallup Workplace Audit as cited by Harter, Schmidt and Hayes (2002).

[h] Survey of Perceived Organizational Support (Eisenberger et al., 1986).

Inter-item correlations are shown in Tables A.5 to A.8 in Appendix 4.

The results per case for all the scale items were combined for each construct to derive total values, thus creating new variables: TotalTPS, TotalLI, TotalOL, TotalWE and then subsequently dividing by the number of items in each scale to obtain values for each construct – TPS, LI OL and WE – within the range of the original scales (1–5).

The Shapiro-Wilk's W tests of normality suggested a departure from normality for some of the computed variables (see Table A.9 in Appendix 5). However, as indicated by the literature (Hill and Lewicki, 2006, pp. 17–18; Pallant, 2007, p. 62) the visual examination of the relevant histograms, normal curves and Q-Q plots (see Figures A.1 to A.8 in Appendix 5) indicated that the data was reasonably normal for the use of parametric approaches in analyses involving these variables, especially considering that the Shapiro-Wilk's test was intended by the authors as 'a supplement to normal probability plotting and not as a substitute for it' (Shapiro and Wilk, 1965, p. 610). The sample was accordingly considered to be normal across the computed variables above.

Statistical analyses were undertaken to address the following:

- What are the significant relationships between TPS, LI, WE and OL?
- How much of the variance in OL can be explained by TPS, LI and WE? Which of the variables is a better predictor of OL?
- Are the variances in OL that can be explained by TPS, LI and WE confounded by rank, age or time in company?
- Are there differences in TPS, LI, WE and OL based on nationality, age, rank, experience as rating and experience with multinational crew?
- Is there a relationship between nature of contract, mode of employment, income satisfaction and WE?
- Do rank, age and nationality mediate the influence of TPS, LI and WE on OL?
- What influence do ship officers believe they have over shore management safety-related decisions?

These were deemed as relevant to addressing the aims of the research question – how shipping companies as organizations learn from, filter and give credence/ acceptability to differing risk perceptions and how this influences the work culture with special regard to group/team dynamics and individual motivation and the specific research question as indicated in earlier in this chapter.

At this stage, all statistical analyses were carried out using Statistical Package for the Social Sciences (SPSS) version 17.

The first step in the quantitative analyses was to examine the correlations between the variables measuring the constructs team psychological safety (TPS), leader inclusiveness (LI), worker engagement (WE) and organizational learning (OL). The correlations were investigated using Pearson product-moment correlation coefficient tests. Preliminary analysis (examination of scatter plot) showed no violation of the assumptions of normality, linearity and homoscedasticity.

The results, as shown in Table 5.2, indicate medium to high[6] positive correlations between the following variables: TPS, LI, OL, WE, rank grouping, age, age (binned), time in company and time at sea. The variable *rank grouping* groups the individual-level data on rank into two groups – operational and management – across onboard departments. *Age (binned)* groups the continuous variable *age* into three: less than or equal to 30-years-old, 31 to 43-years-old and 44 or more years-old.

The relatively high correlations between time at sea and rank, and age are logically to be expected.

Table 5.2   Means, standard deviations and correlation matrix for variables

		Mean	SD	1	2	3	4	5	6	7	8
1	TPS	3.954	.5007								
2	LI	3.851	.5234	0.599**							
3	OL	3.838	.6117	0.626**	0.559**						
4	WE	3.713	.5610	0.564**	0.682**	0.738**					
5	Rank grouping[a]	0.520	0.502	0.394**	0.374**	0.331**	0.330**				
6	Age grouping[b]	1.980	0.818	0.423**	0.385**	0.495**	0.396**	0.568**			
7	Age[c]	37.020	11.338	0.383**	0.339**	0.479**	0.364**	0.521**	0.940**		
8	Time in company[d]	4.682	5.952	0.329**	0.324**	0.289**	0.298**	0.379**	0.419**	0.464**	
9	Time at sea[e]	13.215	10.338	0.485**	0.45++8**	0.472**	0.407**	0.640**	0.838**	0.894**	0.558**

[a] This is a categorical variable: 0 = Operational level, 1 = Management level.

[b] This is a categorical variable: 1 = less than or equal to 30-years-old, 2 = 31 to 43-years-old, 3 = 44 or more years-old.

[c] Correlations for this variable are for the squared transformation of age in years.

[d] Correlations for this variable are for the logarithmic transformation of number of years in company.

[e] Correlations for this variable are for the squared transformation of number of years at sea.

** *p.* < *0.001 lev*el (2-tailed).

---

6   Interpretations of *r*: small (r = 0.10 to 0.29); medium (r = 0.30 to 0.49) and large (r = 0.50 to 1.0) (Cohen, 1988 as cited by Pallant, 2007, p. 132). As Dewberry notes, citing Cohen and Cohen, 1975, 'in organizational research, and social science research generally, a small effect is viewed as a correlation of about 0.1, a medium effect is a correlation of about 0.3, and a large effect would be a correlation of 0.5 or more' (Dewberry, 2004, p. 47).

Statistical Analyses: Amount of Variance in OL that Can Be Explained by TPS, LI and WE

How much of the variance in organizational learning (dependent variable) is explained by the constructs, team psychological safety, leader inclusiveness and worker engagement as independent variables? To test this and find out the relative importance of these variables as predictors of organizational learning, a **standard (simultaneous) multiple regression** was conducted.[7] All the variables met the assumptions of normality as indicated by an examination of the normality plots generated (see Figures A.1 to A.8 in Appendix 5). Evaluations of the results of the regression analysis indicated no violation of the assumptions of normality, multicollinearity, linearity and homoscedasticity. Tables A.10, A.11 and A.12 in Appendix 6 shows the results of the analyses; unstandardized regression coefficients (B) and intercept, standardized regression coefficients (β), the semi-partial correlations (R^2), and adjusted R^2 as well as confidence levels for B, collinearity statistics and F value.

The model explains 61 per cent of the variance (R^2 and adjusted $R^2 = 60$ per cent) – in respondents' perspectives of organizational learning, $F (3,115) = 59.79$, p. $< .001$. This implies that the model as a whole is a significant predictor of organizational learning in shipping companies (as perceived by the respondents). Between the three independent variables, worker engagement is the highest predictor of organizational learning accounting for 57.4 per cent of the variance (standardized coefficient $\beta = .574$) with a positive relation with OL. Team psychological safety accounts for 31.5 per cent (standardized coefficient $\beta = .315$) of the variance with a positive association with organizational learning. The predictive ability of leader inclusiveness for organizational learning was not significant at p. $< .001$. This does not mean that the variable is of no practical significance (importance) but suggests a high degree of shared variance with the other independent variables, team psychological safety and worker engagement.

Statistical Analyses: TPS and WE as Predictors of OL While Controlling For Rank, Age and Time in Company

A second regression analysis, this time **hierarchical (sequential) multiple regression**, was carried out to assess significant relationships existing between TPS, WE (as independent variables)[8] and OL (as dependent variable) while controlling for rank, age and time in company. In other words is the predicting value of TPS and WE for OL confounded by rank, age or time in company? Exploration of assumptions of normality led to statistical transformations[9] to reduce skewness and improve normality, linearity and homoscedasticity in the

---

7   Using SPSS 17 Regression – Linear.

8   LI is excluded since its predictive ability for organizational learning was not significant at p. $< .001$.

9   Using SPSS 17 Transform – Recode Into Different Variables.

age and *years in company* variables.[10] Mahalanobis distance checks for all cases for all the independent variables ($p < .001$) showed no outliers. The resulting square root transformation of age in years, the logarithmic transformation of years in company, and rank (treated as an ordinal variable: $0 =$ Operational level, $1 =$ Management level) were entered in step one. In step two, TPS and WE were entered into the model.[11]

Tables A.13, A.14 and A.15 in Appendix 6 show the results of this analysis for unstandardized regression coefficients (B) and intercept, standardized regression coefficients (β), the semi-partial correlations (R^2), and adjusted R^2 as well as confidence levels for B and collinearity statistics, change statistics and variance analysis for the five variables (three in step one and two added in step two).

After entering *rank* (grouped), *age* (square root) and *time in company* (log) in Step 1 the model explained 24.2 per cent of the variance (R^2 and adjusted $R^2 = 22.2$ per cent*)*. With TPS and WE input in step two, the model as a whole predicted 64.3 per cent of the variance (R^2 and adjusted $R^2 = 62.7$ per cent) in OL, $F (5, 112) = 40.29$, p. $< .001$. For the complete second model only three of the variables were statistically significant: *Age* (square root) at p. $= .002$ and TPS and WE at p. $< .001$. In the final model, TPS predicts 27.6 per cent of variance (standardized coefficient $\beta = .276$) and WE 53 per cent (standardized coefficient $\beta = .530$). *Age* (square root) accounts for 23 per cent (standardized coefficient $\beta = .230$). The implication of this pattern of results is that rank and length of time in a particular company does not improve prediction of respondents' perception of organizational learning. This shows that TPS and WE are still robust predictors of perceptions of organizational learning irrespective of respondent's rank and time spent with company. Organizational learning perspectives were not influenced by rank and/or time spent in company. Age contributes moderately to the prediction of perceptions of organizational learning.

Statistical Analyses: Differences in TPS, LI, WE and OL Based on Nationality, Rank, Age, Experience as Rating and Experience with Multinational Crew

These analyses were done to check for differences between the categorical respondent variable nationality, grouped as OECD/non-OECD and rank, grouped as operational level/management level for the continuous variables TPS, LI, WE and OL. Tests for normality of the grouped samples as per the continuous variables indicated the assumptions of normality for parametric tests were not met. The **Mann-Whitney U test** (non-parametric equivalent of the t-test for independent samples) was therefore used.[12]

Effect sizes were calculated using the formula:

---

10   See Figures A.9 to A.12 in Appendix 5 for normality curves and Q-Q plots of the two resulting transformed variables.

11   From SPSS 17 Regression – Linear.

12   SPSS 17 Analyse – nonparametric tests – two independent samples.

$r = z$ / square root of N

where N = total number of cases (Pallant, 2007, p. 223).
Detailed results of all the Mann-Whitney tests are presented in Tables A.16, A.17, A.18 and A.19 in Appendix 6. Graphical representations of the differences in medians are shown in Figures A.13, A.14, A.15 and A.16 in the same appendix.

The tests revealed statistically significant differences in the following:

- TPS levels for different nationality groups:
  OECD (median = 4.22, n = 28) and non-OECD (median = 3.89, n = 93), U = 820, z = – 2.97, p. = .003 and effect size of r = 0.27 (medium).
  OECD officers had TPS levels higher than those of non-OECD officers. The difference was statistically significant at p. = .003 (two-tailed).
- TPS levels for different rank groups:
  Operational level (median = 3.78, n = 58) and Management level (median = 4.11, n = 62), U = 1012.5, z = – 4.14, p. < .001 and effect size of r = 0.38 (medium to large).
  Management level officers had TPS levels higher than those of Operational level officers. The difference was statistically significant at p. < .001 (two-tailed).
- LI levels for different rank groups:
  Operational level (median = 3.67, n = 56) and Management level (median = 4.00, n = 62), U = 986, z = – 4.06, p. < .001 and effect size of r = 0.37 (medium to large).
  Management level officers had LI levels higher than those of Operational level officers. The difference was statistically significant at p. < .001 (two-tailed).
- WE levels for different rank groups:
  Operational level (median = 3.46, n = 58) and Management level (median = 3.92, n = 61), U = 1118.5, z = – 3.47, p. = .001 and effect size of r = 0.32 (medium).
  Management level officers had WE levels higher than those of Operational level officers. The difference was statistically significant at p. = .001 (two-tailed).
- OL levels for different rank groups:
  Operational level (median = 3.73, n = 57) and Management level (median = 4.00, n = 62), U = 1133.5, z = – 3.38, p. = .001 and effect size of r = 0.31 (medium).
  Management level officers had OL perception levels higher than those of Operational level officers. The difference was statistically significant at p. = .001 (two-tailed).

Differences in nationality scores on the LI, WE and OL scales were statistically non-significant.

With respect to scores on TPS, LI, WE or OL, there were no statistically significant results for differences between those who had experience with multinational crews and those who did not. This was also the case between those who had experience as ratings and those who did not.

In a procedure similar to the Mann-Whitney U test, but this time checking for group differences in the categorical variable age (grouped as less than or equal to 30-years-old, 31 to 43-years-old and 44 or more years-old) for TPS, LI, WE and OL, a **Kruskal-Wallis test** was conducted.[13] There were statistically significant differences between the age groups for all four continuous variables:

- For Team Psychological Safety (TPS):
  Group 1: < 30-years-old (n = 42, median = 3.667)
  Group 2: 31 to 43-years-old (n = 41, median = 4.111)
  Group 3: 44 or more years-old (n = 39, median = 4.111)
  χ^2 (2, n = 122) = 22.159, p. < .001
- For Leader Inclusiveness (LI):
  Group 1: < 30-years-old (n = 40, median = 3.556)
  Group 2: 31 to 43-years-old (n = 41, median = 4.000)
  Group 3: 44 or more years-old (n = 39, median = 4.000)
  χ^2 (2, n = 120) = 22.153, p. < .001
- For Worker Engagement (WE):
  Group 1: < 30-years-old (n = 42, median = 3.333)
  Group 2: 31 to 43-years-old (n = 41, median = 4.000)
  Group 3: 44 or more years-old (n = 38, median = 3.917)
  χ^2 (2, n = 121) = 24.952, p. < .001
- For perceptions of Organizational Learning (OL):
  Group 1: < 30-years-old (n = 41, median = 3.455)
  Group 2: 31 to 43-years-old (n = 41, median = 4.000)
  Group 3: 44 or more years-old (n = 39, median = 4.182)
  χ^2 (2, n = 121) = 34.923, p. < .001

Tables A.20 and A.21 in Appendix 6, show these results:
  For TPS and LI, in general, the older the group the higher the score was.
  For WE, the middle group had the highest score followed by the oldest group.
  For perceptions of OL, in general, the older the group the higher was the score.

Statistical Analyses: Effects of Nature of Employment Contract, Income Satisfaction and Mode of Employment on WE

To examine the effects of the *nature of employment contract* and *income satisfaction* on worker engagement, a **two-way between groups Analysis of**

---

13   SPSS 17 Analyse – Nonparametric Tests – K Independent Samples.

Variance (ANOVA), with post-hoc tests was conducted.[14] *Nature of employment contract* was treated as a categorical/nominal variable (one = Short-term contract renewable; two = Short-term contract not to be renewed and three = Long-term contract permanent), *income satisfaction* as categorical/ordinal variable (zero = No and one = Yes) and WE as a continuous variable.

The interaction effect between *income satisfaction* and *nature of employment contract* was not significant:

F (2, 114) = 0.740, p. = .480

Main effect for *nature of employment contract* was not statistically significant:

F (2, 114) = 1.090, p. = .340

For *income satisfaction* main effect was statistically significant:

F (1, 114) = 15.300, p. < .001

However, even though this variable reached statistical significance, the effect size (partial eta squared) of .12 was small as per Cohen's criteria.

The same test was done replacing *nature of employment contract* with *mode of employment* – a categorical variable (one = through a manning agency, two = directly by shipping company and three = other).

In this second two-way ANOVA test interaction effect between *mode of employment* and *income satisfaction* was not statistically significant:

F (2, 115) = 1.732, p. = .182

Main effect for *mode of employment* was not statistically significant:

F (2, 115) = 0.904, p. = .408

Again for *income satisfaction*, main effect was statistically significant:

F (1, 115) = 9.633, p. = .002

Effect size (partial eta squared) was small (.077) using Cohen's criteria.

The above suggests that, though the variance of *income satisfaction* is statistically significant with respect to worker engagement, one can only make cautious and qualified statements about the practical significance of the interaction between these two variables, since the effect size is so small. Perhaps this is the case because the income ranges of the respondents are such that income

---

14   SPSS 17 Analyse – General Linear Model – Univariate.

satisfaction only marginally explains the variance in worker engagement. There may be a threshold of income below which the main effect of income satisfaction with worker engagement will have much greater effect size.

No statistically significant relationship was found between the mode of employment of a ship officer or the nature of his/her contract and the engagement of that officer.

Statistical Analysis: Comparison of Credence-Giving Factors

One questionnaire item asked respondents to rank what they felt was the importance that their organizations attached to a number of factors when determining the credence to be given to risk information from ship officers. For each factor, no credence was designated by zero and maximum credence by 10.

The factors together with their respective means and standard deviations are shown in Table 5.3.

From this it can be seen that the respondents were of the opinion that the most important factor considered in organizational credence-giving to information from a crew member is rank followed by time the person had spent in the company, the individual's experience with other companies, a person's informal relations with others in the company, age, nationality and gender in that order.

Table 5.3   Means of credence-giving factors

	Credence-giving factors						
	1	**2**	**3**	**4**	**5**	**6**	**7**
	Rank	**Time in company**	**External experience**	**Informal relationships**	**Age**	**Nationality**	**Gender**
Valid N	97	97	96	97	95	96	95
Mean	**7.55**	**6.90**	**6.00**	**5.90**	**5.31**	**5.15**	**3.92**
SD	2.398	2.652	2.608	2.767	2.717	3.312	3.463

Statistical Analysis: Influence of Officer Over Shore Management Decisions

One of the items on the worker engagement scale used was:

What influence do you think you have over shore management's decisions regarding onboard risk? The findings are shown in Table 5.4.

With a mean of 2.76 and standard deviation of 1.354 (n = 121), there is a slight negative skew (see Figure 5.2) indicating that to a moderate to large extent, shore management at least considers the views from the ships. Nevertheless, a substantial 62.4 per cent felt that their opinions were only sometimes *considered* by management or worse.

Table 5.5 shows in tabular form a summary of all the statistical tests and the findings from the quantitative phase of the research.

Table 5.4   Influence of officer over shore management decisions

		Frequency	Per cent
0	Not sure	9	7.4
1	No influence	19	15.6
2	Opinion rarely considered	13	10.7
3	Opinion sometimes considered	35	28.7
4	Opinion always considered	41	33.6
5	Power to change management decision	4	3.3
	Total	121	99.2
	Missing values	1	0.8
	Total	122	100.0

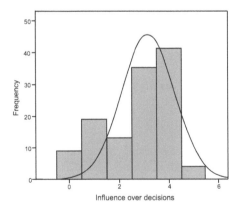

Figure 5.2   Influence of officer over shore management decisions

Table 5.5   Summary of statistical analyses

Research issue	Statistical test	Variables		Remarks
		Independent	Dependent	
Amount of variance in OL that can be explained by TPS, LI and WE	Multiple regression (standard)	TPS, LI and WE	OL	Significance level p. < .001 Model explains 61% of variance WE predicts 57.4% TPS predicts 31.5% LI prediction is not significant at this level
TPS and WE as predictors of OL controlling for rank, age and time in company	Multiple regression (hierarchical)	*Step 1:* Age (square root), years in company (log), rank (grouped) *Step 2:* TPS and WE	OL	After Step 1 the model explained 24.2% of the variance. After Step 2, the model as a whole predicted 64.3% F (5, 112) = 40.29, p. < .001 Age (square root) significant at p. = .002 TPS and WE significant at p. < .001 TPS predicts 27.6% of variance WE predicts 53% of variance Age (square root) predicts 23% of variance Rank and years in company are not significant statistically in the model
Differences in TPS, LI, WE and OL based on nationality, rank, experience as rating and experience with multinational crew	Mann-Whitney U test: Non-parametric test for differences between groups	*Nationality:* OECD/non-OECD *Rank:* Operational/management *Experience with multinational crew:* Yes/No *Experience as rating:* Yes/No	TPS LI WE OL	**Statistically significant differences found in the following:** *TPS levels for different nationality groups:* OECD higher than non-OECD p = .003 and medium effect size = 0.27 *TPS levels for different rank groups:* Management higher than Operational p < .001 and medium to high effect size = 0.38 *LI levels for different rank groups:* Management higher than Operational p < .001 and medium to high effect size = 0.37 *WE levels for different rank groups:* Management higher than Operational p = .001 and medium effect size = 0.32 *OL levels for different rank groups:* Management higher than Operational p = .001 and medium effect size = 0.31 **Non-statistically significant results** *Experience with multinational crew:* *Experience as ratings*
Differences in TPS, LI, WE and OL based on age	Kruskal-Wallis test: Non-parametric test for differences between groups	Age Less than or equal to 30 years/31 to 43 years/44 or more years	TPS LI WE OL	TPS 44 or more years-old (median = 4.111) 31 to 43-years-old (median = 4.111) < 30-years-old (median = 3.667) $\chi2$ (2, n = 122) = 22.159, p. < .001 LI 44 or more years-old (median = 4.000) 31 to 43-years-old (median = 4.000) < 30-years-old (median = 3.556) $\chi2$ (2, n = 120) = 22.153, p. < .001 WE 31 to 43-years-old (median = 4.000) 44 or more years-old (median = 3.917) < 30-years-old (median = 3.333) $\chi2$ (2, n = 121) = 24.952, p. < .001 OL < 30-years-old (median = 3.455) 31 to 43-years-old (median = 4.000) 44 or more years-old (median = 4.182) $\chi2$ (2, n = 121) = 34.923, p. < .001

Table 5.5   *Continued*

Effects of contract nature and income satisfaction on WE	Two-way ANOVA	Employment contract: short-term contract renewable/short-term contract not to be renewed/long-term contract permanent Income satisfaction No/Yes	WE	Interaction effect between 'income satisfaction' and 'nature of employment contract' was not significant F (2, 114) = 0.740, p. = .480 Main effect for 'nature of employment contract' was also not statistically significant F (2, 114) = 1.090, p. = .340 For 'income satisfaction' main effect was statistically significant F (1, 114) = 15.300, p. < .001. Small effect size
Effects of mode of employment and income satisfaction on WE	Two-way ANOVA	Mode of employment through a manning agency/directly by shipping company/ other Income satisfaction No/Yes	WE	Interaction effect between 'mode of employment' and 'income satisfaction' was not significant F (2, 115) = 1.732, p. = .182 Main effect for 'mode of employment' was not statistically significant F (2, 115) = 0.904, p. = .408 Again for 'income satisfaction' main effect was statistically significant F (1, 115) = 9.633, p. = .002. Small effect size
Comparison of credence-giving factors	Descriptives	-	-	Factors for credence-giving in order of importance: Rank (mean = 7.55) Time in company (mean = 6.90) External experience (mean = 6.00) Informal relationships (mean = 5.90) Age (mean = 5.31) Nationality (mean = 5.15) Gender (mean = 3.92)
Influence over decisions	Descriptives	-	-	Opinion always considered – 41 (33.6%) Opinion sometimes considered – 35 (28.7%) No influence – 19 (15.6%) Opinion rarely considered – 13 (10.7%) Not sure – 9 (7.4%) Power to change management decision – 4 (3.3%)

Qualitative Research: Description and Data

The preliminary QDA processes described in Chapter 4 and the associated analyses using Atlas.ti version 6.0, generated 122 initial codes and sub-codes. After the two-stage coding run, some of the codes were subsumed into others, some renamed and others deleted. In the final analysis, there were five thematic codes (with 20 sub-codes), 52 emergent codes and 18 descriptive codes for the interviewees and their companies. A list of the final thematic and emergent codes used is shown in Appendix 8.

The findings from the qualitative phase of the research are indicated in the conceptual maps in Figures 5.3 to 5.6. In the maps the numbers in braces (curvy brackets) show the groundedness and densities associated with each code. *Groundedness* of a code refers to the number of quotations associated with the particular code. It is the first number in the braces. Large numbers indicate strong evidence found for this code. *Density* of a code refers to the number of other codes connected to that particular code. It is the second figure in the braces – following the hyphen. Large numbers can be interpreted as a high degree of theoretical density (Muhr and Friese, 2004).

The figures are representative networks of the qualitative data analysis. They represent causal and influencing links found in the research from the interviews and the document/software perusals.

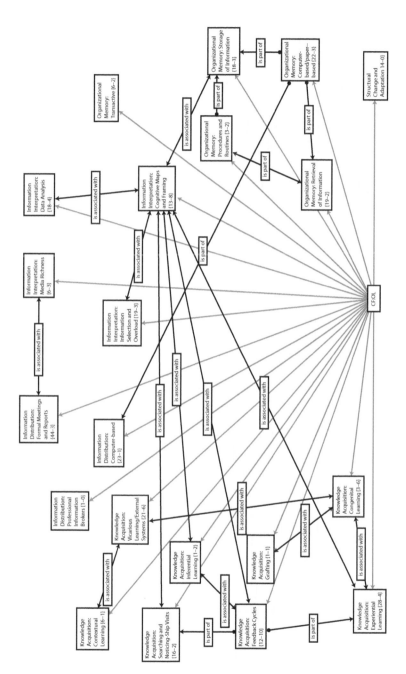

Figure 5.3   Organizational learning: Theoretical-operational themes in network form

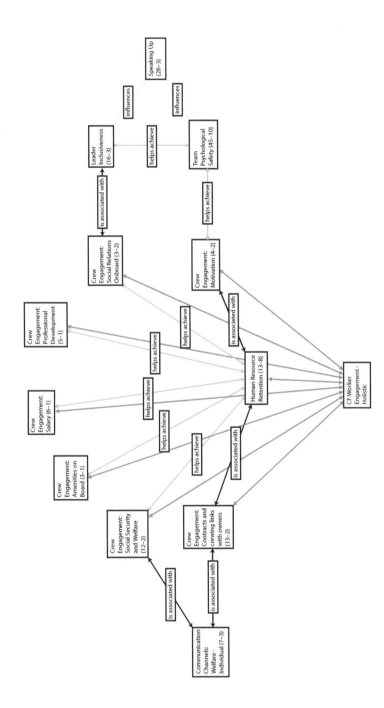

Figure 5.4 Worker engagement: Holistic

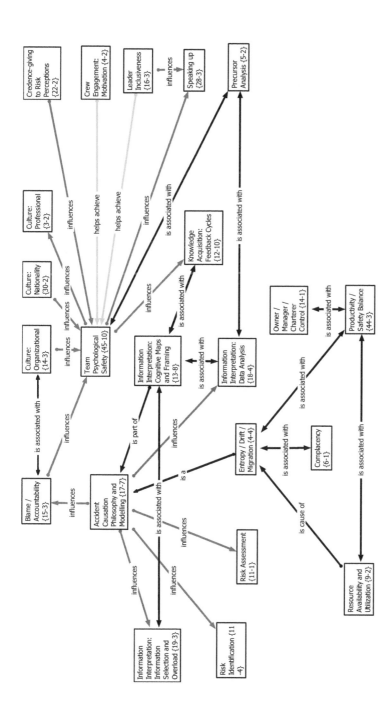

Figure 5.5   Expanded influences in team dynamics

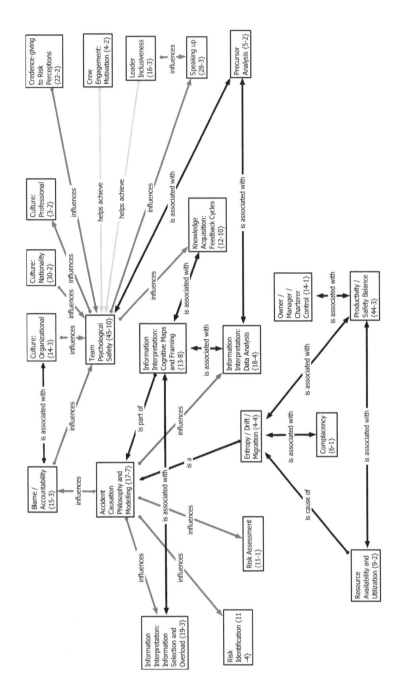

Figure 5.6 Expanded influences in accident causation philosophy and modelling

Chapter Summary

A number of research findings were presented in this chapter. Statistical analyses to answer the research questions were done using the quantitative data from the survey. Findings include:

- Worker engagement is positively correlated with perceptions of organizational learning ($r = 0.74$).
- Worker engagement and team psychological safety are significant predictors of organizational learning.
- Worker engagement levels for management level officers are moderately higher than those for operational level officers.
- Team psychological safety is positively correlated with perceptions of organizational learning ($r = 0.63$), to worker engagement ($r = 0.56$) and to leader inclusiveness ($r = 0.60$).
- Team psychological safety is moderately higher for OECD officers than for non-OECD officers and significantly higher for management level officers than for operational level officers.
- Leader inclusiveness is positively correlated with perceptions of organizational learning ($r = 0.56$).
- Leader inclusiveness is positively correlated with worker engagement ($r = 0.682$).
- Leader inclusiveness is significantly higher for management level officers than for operational level officers.
- Organizational learning perceptions were moderately higher for management level ship officers than for operational level officers.
- Rank and time spent in a company do not predict a ship officer's perception of organizational learning. Age moderately predicts this.
- There are no significant differences between OECD and non-OECD officers in leader inclusiveness, worker engagement or organizational learning scores.
- There are no significant differences between officers with multinational crew experience and those with no such experience on team psychological safety, worker engagement, leader inclusiveness or organizational learning scores.
- There are no significant differences between officers with experience as ratings and those with no such experience on team psychological safety, worker engagement, leader inclusiveness or organizational learning scores.
- Generally speaking the higher the age, the greater is the level of team psychological safety, perceptions of leader inclusiveness and organizational learning. For worker engagement the highest levels were found in the 31–43 year age group and the lowest in those less than 30 years.
- Nature of employment contract (short-term, long-term and so on) was not found to have an effect on worker engagement.

- Mode of employment (whether directly by the owning company or through a manning agency and so on) was not found to have an effect on worker engagement.
- Income satisfaction has a statistically significant but small effect size in respect of worker engagement.
- The most important factor (in the view of the officers) that companies consider when giving credence to risk information from ship officers is rank of the officer. This is followed in order by time spent in company, external experience from other companies, informal relations in the company, age, nationality and gender.
- The majority of officers (65.6 per cent) think that their safety-related opinions are sometimes or always considered by shore-based management or that they have power to change a management decision. A significant 33.7 per cent feel that their opinions are rarely considered, that they have no influence on management or were not sure.

The qualitative analyses show links between the constructs team psychological safety, worker engagement, leader inclusiveness and the theoretical and operational themes of organizational learning – knowledge acquisition, information interpretation and distribution, organizational memory and change/adaptation. Also found were links between these themes/constructs and other emergent themes that influence organizational learning with respect to safety. Predominant among the emergent themes were resource availability and utilization, entropy, drift and migration and accident causation philosophy. The links are shown in network diagrams in this chapter.

A discussion of these analyses/findings follows in the next chapter.

Chapter 6

Discussion of Findings: Theoretical and Operational Themes

We are captives of the moment. The slowness of our thinking and the smallness of information we can process at any one time, our tendency to protect our sense of our competence, the limited inflow capacity of our memory, and our tendency to focus only on immediately pressing problems – these are the simple causes of the mistakes we make in dealing with complex systems.

Dörner[1]

Purpose and Outline

This chapter is the first of two that discuss the research findings. Chapter 6 presents this discussion in light of the literature about the conceptual links between risk and the theoretical and operational themes of organizational learning and in light of interviewees' comments. The themes include processes and structures of knowledge acquisition, information interpretation and distribution, organizational memory and change/adaptation and also levels of learning. The role of the individual/collective is subsumed in the research approach of structuration theory which examines the research data from what individual reports and software/document perusal showed to be recurring organizational practices. Interviewees are quoted extensively in the discussion and in many cases the quotes speak for themselves.

Knowledge Acquisition

Congenital Learning

The primary source of initial knowledge for learning comes from the point at which a company has its inception and who the dominant figures at this stage are. Congenital learning is critical because it sets the tone of the culture of learning that the company then builds on. As Huber (1991, p. 91) writes, 'what an organization knows at its birth will determine what it searches for, what it experiences, and

---

1   From *The logic of failure: Recognizing and avoiding error in complex* situations (1996). Translated by R. Kimber and R. Kimber and published by Metropolitan Books in New York.

how it interprets what it encounters'. A suitable metaphoric conceptualization of this process would be as a 'well-trodden path' that, though patently giving room for deviation, does have a pervasive influence in keeping free-willed agents conformed to its 'boundaries'. Additionally Huber draws attention to the role in subsequent learning of the vicarious knowledge that the founders of an organization incorporate into their approach to organizational work and systems (grafting). They tap into existing societal and industrial norms (professional cultures) and experiences of others (organizational cultures) and use that as quasi-constraining variables in the choices they make about the strategic direction of the new company. Congenital learning is also relevant when companies are offshoots of existing or defunct companies or when one or more companies are grafted together in mergers and acquisitions. The blending of organizational cultures in mergers and acquisitions is often a difficult phase where one or more cultures clash. These intense learning experiences lead to the development of common operational frames of reference for organizational units and individuals and subsequently remain the foundation on which any further learning – or lack of it – develops. Of the companies studied, Chrestos LNG was the closest in time to this stage of learning. This young ship management company (four years at the time of the interview) was obviously drawing substantially from a parent company (Chrestos Hellenic) that had in essence ceased to exist as it was earlier, due to a generational change in leadership and strategic vision. The new vision had led to mergers with other companies. Chrestos Hellenic fleets (dry carriers, VLCC, ULCC, product carriers and chemical tankers) were merged with other companies and only LNG carriers retained under the Chrestos name and management. The reasons for this change in strategy and the explicit focus on the LNG sector were articulated as follows:[2]

Alexander

> First of all, LNG is a very demanding sector. You need your full attention and people dedicated to this business. You cannot run in parallel tankers and bulk carriers and any other business and then focus on LNG. LNG is very, very demanding sector. So we said OK, we are involved in the commercial; we need to establish a new company that will be dedicated to the clients that we have in the LNG sector. The other reason is that this industry is expected that it can be booming – expanding a lot. So we wanted to be in a strategic position, with an organization ready to take challenges and meet chances in the years to come. These were two main factors that led to our establishment of the companies.

---

2   In the main, the language used by interviewees is left unchanged. Apart from changes where there were glaring errors in grammar, etc., the quotes are generally presented as they were stated.

The philosophy of learning and risk perception as well as notions of efficiency in shipping was still very obvious as derivations from that of the earlier company. In the research encounter with Chrestos LNG, repeated reference was made to this drawing on of practices and mindsets from the earlier company. It is noteworthy that though Chrestos LNG is only four-years-old, the language of Alexander draws from a much more extensive temporal span.

Alexander

From a risk assessment point of view, the expertise and experience in shipping gained from many years of existence in generation to generation, as well as the human resources both ashore and onboard, and the procedures we lay down, is the major control measure to associate the risks in all the phases of the business.

In shipping, as in other contexts, congenital learning in companies (as 'progenies' of other companies) was found to be linked to experiential learning that revolves round the memories and practices of social units and individuals who are connected in some way to a 'parent' company – whether existing or defunct or who join the nascent organization with experience from other organizations. The effect of this is critical to the analysis of the dissemination of risk notions and perspectives when shipping companies set up manning agencies, training institutions and similar bodies. Congenital learning suggests that such progenies will be partly restricted in their scope of operational philosophy due to the 'knowledge corridor' created by the parent company. It is appropriate therefore that such links be explored in any assessment of such institutions.

Despite the wider gap in time from inception, the other companies did show signs of how pioneering philosophies and cultures set in national and regional origins hold sway in the development and retention of organizational culture and how they remain a dominant effect even with high employee turn-over through the years. This was sometimes the case even where different corporate entities had specifically been established to take advantage of different operational norms.

Tim

There is also a cultural issue within Alpha Tankers. Bear in mind we are an Asian company. The Asians [specific nationality mentioned] have a very particular way of working. They've had it for a very long time and they are culturally resistant to change. They want to do it their particular way. They feel their particular way works. It works for their culture and – I choose my words carefully – they don't often recognise that their way doesn't necessarily work for other cultures particularly well. That having been said the Magna Group prepositioned companies around the world to take advantage of other countries and other cultures. At a higher level within the organization, I think they recognise this very, very well. I just think of the middle management level within the group of organizations does not translate particularly well. But for example we are based in London. We deal primarily with the western hemisphere, deliberately because we are western, because we understand the culture, because we understand how it works. We have another organization exactly the same as us in another location in Asia. They deal predominantly with the Mid East and Far East, primarily because that's where they are and that's what we do. And it works very, very well. But you have an overriding authority who tries to conglomerate everything and tries to get benefits of, tries to get the benefits of being a large organization and focus them at every level within the organization instead of saying 'OK, this is where we can all benefit; this is where we can't. You manage your organization to fit what you do. You do what you do and we together will work something out from it'. And we have designed those systems so they are common within the Magna Group organization.

In ways such as this, 'restraints' of congenital learning constrain the organizational and learning processes of progeny entities. Similarly where congenital learning is positive, the organization benefits in the long-term from an enhanced approach to learning.

Experiential Learning

Experiential learning appears to be the most dominant form of learning in the shipping companies. By this is meant the intra-organizational learning that takes place by virtue of occurrences on ships and offices of a single organizational entity. It is dominant in the sense that it receives the most attention. The experiences of crew members and objective occurrences on board are conveyed in various ways to decision-makers in the shore offices. These serve as the primary source of risk knowledge and by extension learning.

Paul

The main source for any information on risk is the vessels – definitely: the feedback that we gain from the vessels in forms of nonconformity reports, near accidents reports, accidents that do happen, could be smaller accidents or the larger ones as always often like whenever it happens and then a lot of follow up.

We have the lessons learnt which is the experience-sharing between all the vessels and all the different. We have the PEC [Protection and Environmental Committee] meetings; security meetings or standard meetings that they have on board, the whole crew, to discuss the lessons learnt for the past that we have in our system that are new quarterly reports, all kinds of information that they have received on risk and risk management as well, new procedures. So when these lessons learned are published, for instance, they take that up in the first PEC meeting they have on board. For onboard safety awareness and risk management in that aspect, lessons learnt are probably the most important thing we have. So based on all the nonconformity, near-accident reports, accident reports, Jack [Risk Manager] or someone else will go through the reports; what happened, what do we need to be aware of, what do we need to focus on. So for instance, you have splash accidents for chemicals, get some splash in someone's eye or whatever, and then we focus a lot on safety equipment, safety equipment, this is important now! And they take that up in their meetings onboard.

Edith

Normally we have these incident reports or SAFIR [Safety and Improvement Reporting System or Safety Information Reporting] like near-misses. When you say near-misses, it is this almost an accident but because it was prevented it was a near-miss. That's why we call it a near-miss and by means of reporting and by taking note of all this and collating it in one file, putting it in the system – because we have a system where everyone can view, everyone can read and then those responsible person will answer, the lesson learned and everything like that and then we will collate it.

Tim

We've had a number of occurrences where – a very basic one was a third engineer who was transferring chemicals and wasn't wearing PPE [Personal Protective Equipment] splashed his eyes with chemical. We had to evacuate him ashore and we've investigated that and found that it was to do with one not even wanting to wear the safety gear because it wasn't comfortable. Two, thinking he could get away with it. We've focused on that, pushed on that, pushed very hard on that, to learn, to prevent that reoccurring.

Alexander

Yes [risk information to Chrestos LNG] comes mainly this is the source, through the operational, daily operations [on board]. Mainly I would say that through the daily operations and the environment that we work with. And again I am mainly speaking about technical and the ships' actual operations.

Vasilis

[When an accident occurs] we see what was the problem and this casualty occurred, so we see our risk assessment in place and in case we find I mean measures which were not duly taken, we enrich our, the already existing risk assessment on ship with the controls we missed to implement, in order to avoid the casualty or the accident or the incident which happened. So this is a dynamic process and we learn from what we have suffered, enriching thus our risk assessment system.

This philosophy of learning from the experiences on board ship was a dominant theme with all interviewees expressing the opinion that this was the primary way in which learning was done in shipping. In most cases, experiential learning involved lessons learnt that are direct, objective and in reaction to specific incidents. It results in an enhanced focus on a particular incident or series of incidents. While there are many merits to this, one obvious demerit is the way it relies almost exclusively on a reactive paradigm. During episodes of experiential learning, such learning is confined to specific incidents and there is little evidence of seeking experiential learning from the analysis of patterns and underlying conditions.

Figure 6.1 is a screenshot of one way in which Olive Shipping receives the lessons that are then fed into system procedures as experiential learning.

There are suggestions in the literature that point to 'organizational experimenting' as one of the important aspects of experiential learning in organizations. Experimentation on a large organizational scale is most obvious in organizational learning literature associated with new product design (Edmondson, 2005; Rodan, 2005). There was little evidence of such large scale experimentation in the risk and safety domain in the research carried out. This is perhaps because

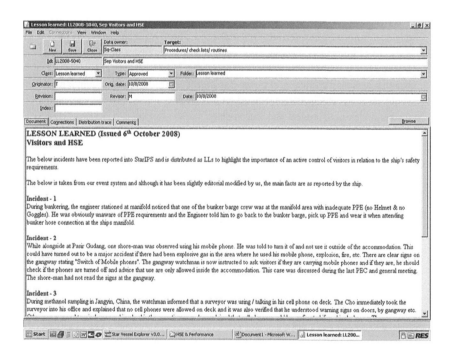

Figure 6.1   Screenshot of experiential learning input: Olive Shipping

using intentional experimenting so as to learn from risk consequences post-event[3] does not seem prudent given the potential negative consequences. Experimentation is essentially a trial-and-error process and in the shipping industry when used, it is in small steps and on a small and often localized scale (see discussion of this kind of experimental learning in Miller, 1996, p. 493).

Paul

> Olive Shipping's strategy has never been to be at the forefront of innovation. We're kind of conservative in that we wait until we see that it is working well which again is a good and a bad thing.

While such experimenting is evident – even necessary – in industries that are involved in new product design, the safety world of high-risk industries does not grant shipping companies the kind of margins that allow for intensive experimenting as basic modus operandi. Organizational experimenting, when it happens, is constrained by careful resource allocation and monitoring (Rodan, 2005).

Organizational self-appraisal and performance monitoring are a key part of experiential learning. Among other methods, internal safety audits, being a requirement of most certificated safety management systems such as the ISM Code[4] and ISO 9001,[5] are carried out extensively by the health, safety and/or quality departments of all the companies researched. Such audits yield copious amounts of information for learning. It is noteworthy that both standards mentioned above do not stipulate a fixed time frame, only using the terms 'periodically' and 'planned intervals'. The companies generate the planned intervals most normally, quarterly, semi-annually or annually and then appear to append their safety review and analysis cycles to these dates. This way of operating (segmentation of risk data) may have negative consequences. The 'calendar-based demarcations' form what Kiesler and Sproull call 'organizational breakpoints' (Kiesler and Sproull, 1982, p. 561) which may introduce salient points that are not necessarily relevant or important to risk management. They argue that salient features in a stream of information affect managers' perceptions of causality, relevance or importance.

---

3   This is not the same as carrying out safety and risk associated drills of predetermined procedures.

4   Paragraph 12.1 of the ISM Code requires that 'the company should carry out internal safety audits to verify whether safety and pollution-prevention activities comply with the safety management system' (IMO 2002).

5   Section 8.2.2 of ISO 9001:2000 requires organizations to 'conduct internal audits at planned intervals to determine whether the quality management system (a) conforms to the planned arrangements, to the requirements of this International Standard and to the quality management system requirements established by the organization, and (b) is effectively implemented and maintained'.

In their words 'managers' perspectives affect their punctuation of segments of behaviour; segmentation then has implications for the manner in which decision-makers frame problems and solutions' (p. 561).

Experiential learning also involves other modes of institutional self-appraisal with, for example the carrying out of crew appraisals as an index for crew competence in risk-related matters. In most cases this is done through the agency of superior officers who comment on the competence, attitudes, and so on, of subordinates. It is noteworthy that a respondent in the questionnaire survey raised the issue of appraisal of senior officers by junior officers:

> I think junior officer are also possible to appraise senior officer (captain). Inter-appraisal like this can bring out safe operation at sea (23-year-old South Korean A/O – Male).

In some cases, learning in the organization was not foreseen or planned for in any formal way. Though all the companies show evidence of some level of system-designed structures for learning, there were also many instances where intuitive or subjective sentiments focused attention on learning issues and areas.

Alexander

> I mean a technical inspector goes on board for a – to repair a part, and then he thinks that the atmosphere or the deck officer and engineer officer are not going well along, or the officers and the crew is totally separated, there is not any … then we also send some more people on board.

In situations like the one of which Alexander speaks, there is no a priori intention to acquire knowledge for learning. In an unintentional and perhaps unsystematic manner, cues are picked up that suggest a knowledge or attitude gap and hence a learning need. The companies then try to respond to this 'discovery' utilizing the existing formal system.

Regarding formal system features that are designed to monitor feedback, Olive Shipping uses an elaborate computerized system that allows for relevant persons to monitor and input into the status of any particular event. Figures 6.2 and 6.3 show screenshots from this system. The areas in the super-imposed boxes show the status of the event (incident/accident/near-miss, and so on), in this case showing 'closed', meaning the safety issue is purportedly addressed at all important levels of change, learning and dissemination of information. The system is designed to bring to the attention of the relevant person(s), any outstanding issues.

All experiential learning is obviously heavily dependent on feedback cycles. In the foregoing, two coexisting approaches to feedback are discernible. On the one hand there is the feedback that is acquired in an unintentional and unsystematic manner and which appears to focus more on precursor conditions, that is, pre-accident behaviours, attitudes and conditions that, in the eyes of the observer,

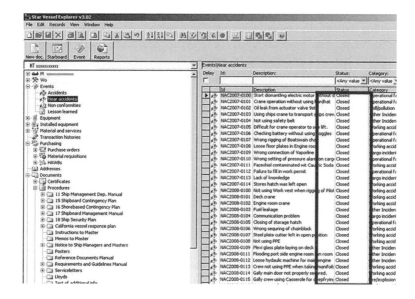

Figure 6.2   Example of computerized feedback system showing event status

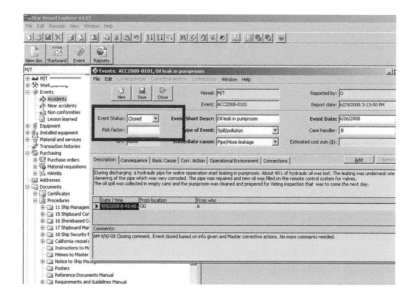

Figure 6.3   Example of computerized feedback system showing event details

could have potentially negative consequences regarding safety. These often subjective views may have no reference to earlier accidents, either because these have never happened, or because of limited awareness of them by the observer. Such feedback may then be assimilated into more structured safety management systems. On the other hand feedback cycles are part of mandated and documented safety systems that are often reactive being essentially a response to incidents (near-misses, accidents, and so on) that have occurred.

The former approach requires what this author is labelling 'inferential learning' and others have called 'inductive learning' (Maier, Prange and Rosenstiel, 2001). It is the kind of learning that uses abstract patterns (as distinguished from the specific and explicit events that are understood to relate to particular accidents) from an organization or individual's understanding of contextual issues, to infer patterns of causation in the present or future. It suggests the use and exploration of potential scenarios that extrapolate from these patterns. Galer and van der Heijden (2001, pp. 849–50) indicate how some organizations have used patterns in the form of frameworks or project scenarios in management and business settings. Citing Loasby (1992) they define such scenarios as 'coherent sets (or sequences) of possibilities to the realization of which no fatal objective is perceived'. The following is adapted for risk from what was originally a quote addressed to business:

> Scenario planning is a way to learn ... through seeking, in an informed and disciplined way, to map the many different ways in which [the future] could evolve. It develops and uses people's understanding of underlying structures and driving forces in order to help identify possible future discontinuities and other significant changes in the *risk* [original word here was *business*] environment. (Galer and van der Heijden, 2001, p. 852)

Sometimes such processes of learning have to be based on samples of less than one (March, Sproull and Tamuz, 1991), relying for their abstraction on similar event dynamics and not specific sameness.

In recognition of the importance of acquisition of knowledge in both the formal and informal feedback ways, all the companies put a heavy emphasis on ship visits by shore management. While technical people, such as vessel superintendents, are required in most cases to visit the ships, some companies went further with a requirement for all senior managers (not necessarily technical) to visit the ships regularly. This is particularly true for the tanker-owning companies where such visits are a requirement under the Tanker Management and Self Assessment (TMSA) scheme originated by the Oil Companies International Marine Forum (OCIMF), currently in its second edition (Oil Companies International Marine Forum [OCIMF], 2008). Such senior officers' visits to the ships are primarily a scanning or familiarization mechanism for operational and safety information, but also is intended to help create a comfort zone with the crew to encourage the uninhibited disclosure of risk information.

Evan

So how do we identify risks? I mean a good example is, you know visits on board the vessels, you know where … again we have a requirement to visit the vessels. The technical superintendents will visit the vessels up to three times a year. That's mandated and never, never more than six months should elapse. The operations superintendents will visit the vessels a minimum of once a year … We also have a requirement from the Magna Group head office that senior management should visit the vessels. So myself and the general manager of our organization, you know, try to visit the vessels on a regular basis … We visit our vessels a lot, you know, and obviously we are on lookout. We're on the lookout to try to communicate that, you know. We communicate that [the company is open and welcomes all risk information and perspectives] … You know regular ship visits are very important OK. And when you go to visit a ship, you don't want to meet just with the, you know, with the captain and the chief engineer. When I go to visit the ship I always try to arrange a meeting with just the crew and myself OK.

Alexander

We sail with the vessel … and carry out not an official but on the job assessment and the structures, etc., etc., have discussions with everybody on board.

The ship visits have other advantages:

Jason

What we've done here I think is starting to work. I mean people like myself, like the general manager, the managing directors, we all go to the ships as well and we get our face known to the people and the people, I think they appreciate that. For years none of the senior management went to the ships and I think we are one of the few companies that is actually doing it. And I think it's helped a lot. They know who they are speaking to there, you know, they can put a face to, you know what I mean, and I think that, and I think I would say in the industry they just don't do that elsewhere. I think with Alpha Tankers anyway, I think we've been quite progressive in getting senior management out to the ships, showing that we're committed to the company and that I think the people appreciate that, you know

Tim

We have inspection routines where all of our superintendents will visit our ships two or three times a year and look for these factors. Because we have what we call 'high focus areas' which are areas where we have had something go on that we think needs attention and the ships need to pay attention. So we've got items which are big hit one events – yeah perhaps we can't do trending on or we don't think are indicative of a trend – we can put them into the high focus areas list and our superintendents when they visit the vessels can focus on those areas when they go on board and we could see from those visits whether they are a trending item or whether they're a one-off. If they're one-offs well then we were unlucky; if they are a trending item, then we're focusing on them, we're improving, we're setting the focus for all the ships when we visit the ships.

27-year-old 2nd Asst. Engineer – Male

Shipping companies often send representatives/technical superintendents to their ships. These personnel assess ship's condition and part of their job is to talk with ship's crew and collaborate ideas, etc.

There is an obvious assumption that people who visit the ships (either as a safety management system requirement or informally) are equipped to appreciate the differences of context, what constitutes normality in safe ship operation and crew behaviour as well as an understanding of the learning processes involved. While this may be so in specific cases, it is not an assumption that should be generalized. There appears to be limited training in METI (Maritime Education and Training Institutions) or other educational contexts (where most superintendents and managers are drawn from) that prepares individuals for this kind of ongoing iterative inquiry in the maritime sector. The industry as a whole assumes that technical expertise and competence augmented by familiarity with shipboard life – no matter how removed in time from the present – is sufficient for this 'double-loop' learning mentality. This kind of mentality focuses on inquiry that goes beyond just the welcoming of 'suggestions'. It implies active searching and noticing, a scanning approach that emphasizes practical measures to solicit risk information and that go beyond simple statements of 'encouragement'. In this regard the importance of 'risk awareness' and optimum and open communication between all levels of the safety management structure cannot be overemphasized.

Vicarious Learning

Another form of learning that requires feedback is vicarious learning. This refers to the learning that organizations undertake based on events in other contexts for example, other shipping companies. This type of learning is given some attention but not as obviously systematic as in experiential learning. The majority of vicarious learning opportunities come through the agency of third parties such as commercial customers, the insurance industry, classification societies and trade organizations. At one level the risk information that comes in through classification societies and insurance companies was contextualized and rapidly disseminated, sometimes with procedural and structural changes. At another level, generic risk information in the industry is seldom captured in a predetermined procedural manner and as part of a safety management system. The approach was one of unconscious organizational assimilation by virtue of individuals in the organization being exposed to such information on risk, for example through publications of trade organizations. For the tanker industry, the commercial interests (shippers) have a key role in the dissemination of risk information and the tanker shipping companies placed a significant amount of emphasis on this as compared to the dry trades. Commercial motivation for enhanced safety is by no means insignificant.[6]

Paul

> We have the oil majors on our back all the time vetting and the vetting regime they have is a large part of the risk management information as well. Recently we had some comments from Exxon Mobil which made, that we needed to change the task risk assessment ... Other sources [of risk information] could also be, you know those circulars that go from the insurance companies, US Coast Guard, Veritas. So all these things help us identify risks. You have both the, you have the insurance companies, the oil majors, you have several sharing programmes within that, sharing, you know near accidents, accident reports, especially the oil majors have quite a few issues that come by.

Jason

> I mean we have ... like everyday actually we will get something from Asia like, they have these like, it used to be Tankers Supervisory Group and what they used to do is like they send out these advisory notices that ... problems on ships. They also have all the shipping notices and stuff like that ... special advisories like that they send out to us as well. We also have anything that comes in from Class, you know from Class NK[7] or DNV or we have these sent out to the ships as well. A lot of them are integrated into SHIPNET now [SHIPNET is the computerised database and safety system being used by most of the companies in the Magna Group]. We use to send out separately but now we are actually importing them into SHIPNET and all the ships are automatically receiving these updates. You know introduced from Class.

The Magna group companies were unique in that they try to learn vicariously at different levels. First they have vicarious learning opportunities at the within-Magna level and then also at the level of the completely external shipping industry. It appears this was not always the case with the group. In cognizance of a historical phase in which the group as a whole was 'rudely awakened' from a period of

---

6   See Section 7.8 on resource availability and utilization, organizational entropy and drift.

7   NK – Class Nippon Kaiji Kyokai – Japanese ship classification society.

complacency, emphasis is now placed on the creation of verifiable systems of inter-company learning (within the group). This seems to have had a knock-on effect in also bringing learning from other external companies to the fore.

Tim

> We were all separate companies – completely separate companies. We had 25, 30 ships here and the same in the location in Asia and never the twain shall meet. We did not communicate with each other; we did not work with each other at all. Now we do and now we learn from each other and get more out of it and actually work together much better that we ever did before ... From an external learning perspective, yes, we have. Twice a year we attend the International Tanker Forum [ITOSF – Informal Tanker Operators Safety Forum]. The International Tanker Safety Forum was last week in Singapore. In fact my colleague was travelling there and we take things away from there very often. We are part of a discussion group that go on with those. We're members of INTERTANKO [International Association of Independent Tanker Owners] so we get their bulletins, their information as well. And things like this Root Cause Analysis Grid. We didn't develop that. That came from other people.

Vasilis

> [We also learn] from other industry, from the safety bulletins of other companies, from various reports we receive from oil majors. [We benchmark with] other companies. And also there is an informal safety officers meeting [ITOSF – Informal Tanker Operators Safety Forum] which is being done, I think three times per year in various places of the world and safety officers of the companies, I mean the safety officers of the companies, they are gathered, they are exchanging ideas, they're benchmarking by presenting the accidents, near-misses. All these things. So they have, I can have the LTIF [Lost Time Injury Frequency] of another company, I can have total quarterly cases of other companies so I compare them, I compare their let's say performance indicators together.

One necessary extension of vicarious learning should be the *contextualization of learning*. Anecdotes abound of organizational mimicry that bordered on the ridiculous because of the failure to contextualize events external to an organization's own environment. For example, when the ISM was developed, some good companies formalized excellent systems that were already tacitly in place. Some of these procedures for optimum safety management systems ended up on the market, where they were 'snapped up' by other less-informed companies and sometimes used without any notion of adaptation for the particular context. As Anderson observes (2008, p. 14), 'you can find yourself [as an auditor examining the standard procedures of a company] asking a company how it intends to carry out crude oil washing on board its ferries!' Imitation or mimicry is particularly relevant in a high-risk industry where it may not be prudent to engage in experimentation as a learning process. The disadvantage is that it often precludes the specific contextual parameters in which the solutions have been applied in the model organization. It is necessary that the contextual fundamentals are applied to avoid necessary imitation (or inter-organizational learning) becoming a *de facto* experimental process.

The contextualization of vicarious learning is important and blind mimicry is generally not encouraged by the companies in this sample. The example Jason gives is typical.

Jason

If it is relevant to our ships then yes [it is analyzed and incorporated into procedures]. If it's not, it might be something completely different because I mean, for example say if it is something like a medium speed engine on a car carrier or something like that, I mean we don't have any medium speed engines. They are all slow speed diesel.

Vasilis

There is always something new; there is always something which is, let's say, which is not generic [and needs unique application and contextualization].

The factors to be used for contextualization should be carefully considered. In the quote below, a respondent speaks of contextualizing (comparing) based on fleet size which in the safety and risk context, unnecessarily restricts the potential scope for vicarious learning opportunities.

A respondent

So you can make comparisons. And of course we are talking of companies which have the same number of vessels. I cannot go to compare myself with a company which has only four vessels. I will go with a company which has around the same number of vessels – for example 40 vessels. OK I cannot compare myself with a company which has only dry cargo vessels; something different OK.

In this last quote the basis of contextualization is number of vessel. It certainly is possible that a company with a large number of vessels can still learn from one with only one, provided the determinants of a particular safety issue are held in common, such as type of ship and so on. Arguably number of ships should not be a parameter for benchmarking with regards to safety. The presence of this view shows how the selection of issues to use for vicarious learning is subjective and limited to the perspectives of key agents in organizations.

Different forms of vicarious learning were found in the research – such as between their own ships or between company departments – but inter-organizational vicarious learning was more restricted as compared to experiential learning. In all cases this aspect of vicarious learning was evident in only a limited way with the company systems capturing these to different degrees. It appears, however, that a significant amount of learning exists outside the scope of formal systems and is intuitive and informal suggesting a high degree of subjectivity in risk information selection, safety choices and by extension what is considered important for training.

In the context of lessons to be learnt vicariously from the industry at large, one notes the IMO's attempts at having an industry-wide database of safety-related incidents for the purpose of learning lessons – the Global Integrated Shipping Information System (GISIS) and the Flag State Implementation (FSI) Committee's lesson learnt circulars. At more localized levels, the maritime industry has some reporting systems that also work to enhance vicarious learning. They include the Nautical Institute's Marine Accident Reporting Scheme (MARS)[8] and the UK's Confidential Human factors Incident Reporting Programme (CHIRP)[9] for general aviation and the maritime industry. Kristiansen (2005, p. 420) reports on similar systems for Norway (MARINTEK) and, Finland (VTT). Sweden's *Informationssystemet om incidenter i sjöfarten* (database of incidents in shipping) – INSJÖ – may be added to the list. At a pan-national maritime administration level, there is also the European Marine Casualty Information Platform (EMCIP) which is under development and is more focused on accident/casualty investigation reporting. Undeniably all such databases are extremely helpful in risk analysis. They are dependent for their success, however, on the degree to which they are known about, used, and share common classification schemes as well as the legal protections they afford in terms of jurisdictional scope. Good examples of such programmes from other industries are the US Aviation Safety Reporting System (ASRS)[10] run under the auspices of NASA (the US National Aeronautics and Space Administration) and the Process Safety Incident Database run by the Center for Chemical Process Safety (CCPS).[11] The ASRS is (crucially) operated by a long-term contractor with focus solely on this service, and provides legal clauses that allow for higher levels of feedback (Tamuz, 2004). It is unbiased/independent and legally qualified by way of immunities and clemency clauses. Reports come into the ASRS database from all areas of the industry: from air-traffic controllers, pilots and their flight crews, mechanics and maintenance crew and all other aviation professionals. The benefits of such external incident databases are significant and clear. They include better accident, incident and precursor trend analyses because

---

8   http://www.nautinst.org/MARS/index.htm retrieved 22 July 2009.

9   http://www.chirp.co.uk retrieved 22 July 2009.

10   See http://asrs.arc.nasa.gov retrieved 21 April 2009.

11   See http://www.aiche.org/CCPS/ActiveProjects/PSID/index.aspx retrieved 21 April 2009.

of higher levels of statistical data, greater opportunities for vicarious learning, increased exposure to accident precursors in the wider industry and valuable data for training purposes (Kelly and Clancy, 2001).

In the research visit to the companies there was only one respondent who made reference to the IMO circulars and such databases.

Tim

And we also have industry wide practices. IMO circulars, Safety at Sea International, Seaways Magazine from the Nautical Institute, CHIRP, Alert; all of those periodicals you get coming out that tell people what's been going on out there.

While it may be the case that the companies are aware of the existence of such databases, it appears that they are not given the necessary attention by the companies. It is possible that this is because there is not enough 'public awareness' (with specific relevance to the necessary target groups) of the value of these services or that they do not address the needs of the companies.

In summary knowledge acquisition as part of the organizational learning theoretical model is manifested in shipping companies through congenital, experiential, and vicarious learning with a degree of cognizance of context and the acknowledgement of the importance of feedback cycles. In keeping (purportedly) with the requirements of ISM (quote), all the companies visited make use of documentation of procedures. No company had structured procedures indicating systems for all the specific modes of knowledge acquisition as indicated earlier in Figure 3.2. The component most significantly captured by the companies in the knowledge acquisition step was risk information arising from experiential learning – either from accidents or from near-misses – for the particular company. Experiential learning is still restricted to accidents and risks that have been recognized in the industry for decades: personal injuries, collisions, groundings, fires and so on. The pattern of learning is evidently a single-loop one with no questioning of fundamental and generic ship operation traditions. This makes the experiential learning that the shipping industry engages in very similar in learning content to vicarious learning. It is more a reinforcement of what is common knowledge by way of potential for accident causation, rather that new knowledge that shifts basic paradigms.

Information Selection and Interpretation

To select risk information the agents of the companies (as individuals and groups) determine relevance, value and therefore what to give credence to based on their own cognitive maps and frames of reference as influenced by their backgrounds and the organizational culture. The companies resort to the use of computerized systems to track the information and to help with the analysis, but it is essentially a subjective assessment of risk level based on pre-designed risk matrices.

In the following quote, Paul describes input into computerized elements of Olive Shipping's safety management system.

Paul

The way it works is that in any one of these databases, you have an 'event status'. So you have an event status, so a new event will go in. If it is a new one then we have fleet coordinators. Fleet coordinators will go in and find every new case that has come up that is new in the system and will go to the fleet managers. So the fleet managers know what cases they need to follow up. Then it is up to the fleet managers, in coordination with marine superintendents, and the risk management department, Jack, and the quality management to look at these cases individually, all of them, to see what is wrong, to make recommendations to what could be done to prevent this, how do you solve it, what would be the advice for their fleet.

In recognition of the existence of a potential for biased subjectivity, the companies try to create common cognitive maps and frames of reference for their employees regarding risk.

Alexander

By this model [referring to the accident causation model that Chrestos uses] everybody can have the same picture.

It is debatable whether this is successful or whether a high level of commonality in perspectives is advisable. It raises the issue of groupthink with the attendant loss of critical analysis and the quietening of dissenting views which may hold great value.

Another part of this step of organizational learning is how the data that is selected is analyzed and interpreted. In Olive Shipping a system known as STAR IPS (Star Information and Planning System) is used in the analysis of risk information. It incorporates reports on safety known as Star Event Reports.

Paul

Star Event Report will be reports based on the event system here. It's basically just summaries that could be drawn out from the information that has been put in the accidents, near accident reports and so on. We have reports. You can pull out the consequence analysis, accident analysis: the system will take anything. Instead of 'near-misses' we use 'near-accidents' ... Yes there is a system. I will show you in STAR IPS because, I mean, basically the vessels use STAR IPS to report nonconformities, near-accidents, accidents and then these are kind of left open on the system. Now because they are open they are going to be dealt with by our shore personnel as well. So they'll be looked at by the fleet managers, consulting with the marine superintendents and the marine managers and so every report that is sent, has to be answered and closed.

The reporting and analysis system is quite versatile and the company ensures that it undergoes consistent review. It was probably the most advanced data management system in all the companies visited and certainly will represent a not too common data analysis tool in the industry considering that the sampled organizations are higher-end operators. Olive Shipping admits that incorporating the computerized system was not easy but a focus on safety, they say, made it an ongoing task to which the organization gives priority. An earlier computerized system for monitoring safety called RISKBASE (see Figure 6.4), was first-generational in character and could not meet the optimum demands of contemporary risk data analysis. Figure 6.5 shows the newer system.

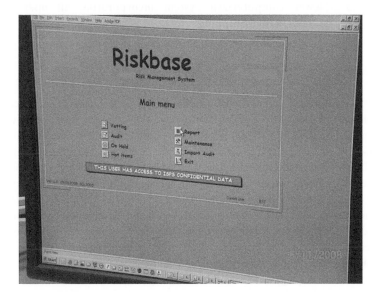

Figure 6.4   Riskbase: Old computerized system used by Olive Shipping

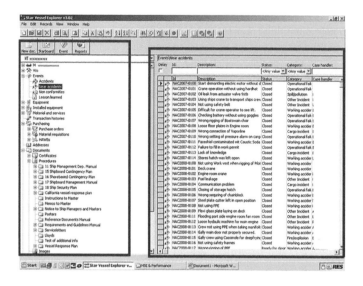

Figure 6.5   Newer computerized system incorporating all operational systems: Olive Shipping

Barry

We have struggled a lot. It has been hard work and we are not there yet to get 100 per cent because it is not very easy to agree with some data freaks how we want it [to reflect data]. STAR IPS is not in-house. It is, let's say we have an in-house tool here which is taking care of the … this database ['RiskBase']. RISKBASE is an in-house tool. This is taking care of all the vetting … this is the SIRE [Ship Inspection Report Programme] and CDI [Chemical Distributors Institute] inspection, where we have our thing here. We can take reports and we can distribute, which we send to ships and everything. And this one is something we will just dispose of – throw away. The thing is why we are throwing it away because this is based on simple, and old, it's only an access database. So we are transferring this one to the new system … So we have systems which we are working, developing all the time. It is an ongoing process.

Almost 10 years after the coming into force of the ISM Code for all ships it was noteworthy that even in these leading-edge companies, highly versatile and able systems for fundamental risk data analyses are only just evolving. Data input and analyses are still very much text-based and essentially descriptive – and therefore reactive – in nature.

Tim

We actually do a break down of analysis which is more important from a management perspective ... Prior to our setting up and running the ship management system, which is something that we've done in the last 12 months (I spearheaded the project on this and I've got it up and running in the last 12 months); prior to that we had a paper and electronic based system. So we had a system where we recorded and stored it on our network and you could go back through and review them. And we did, we still did annual analysis on it and we still used the same framework to do that analysis that we are using now. And what we actually do is, we do a quarterly and an annual review of all of our occurrences. Now an occurrence is accident, near-miss, technical incident, failure or something or other where nobody got hurt, basically piece of the ship or piece of equipment failed outside of the routine planned maintenance cycle, audit inspections, all of those are put together and reviewed and we use the same root cause analysis for all of them and apply that root cause analysis across everything. So last year we had 2170 odd occurrences across the board so we can now see, with that kind of number you can see that we get a real actual – a statistic that's worthy; a statistic that's believable. If you've got half a dozen accidents through the year, that's not reliable data. It doesn't show you trending. It shows you points. But if you've got over hundred or over two hundred or over, in this case two thousand, we've now got a reliable statistical pool that we can pull from and actually learn from it and benefit from it. So what we do is we have this annual review.

[With regard to those incidents that don't form part of a larger trend but are serious]

you look at everything that is going on around you. We will do trend analysis on those items and we would always look at them and say is this indicative? ... Because we have what we call 'high focus areas' which are areas where we have had something go on that we think need attention and the ships need to pay attention. So we've got items which are big hit one-off events. Yeah perhaps we can't do trending on or we don't think are indicative of a trend – we can put them into the high focus areas list and our superintendents when they visit the vessels can focus on those areas when they go on board and we could see from those visits whether they are a trending item or whether they're a one off. So we might not be able to demonstrate a trending, but we are also not ignoring them.

Alexander

We also analyse them and mainly we focus on the trend. It's not that we don't care how many near-misses we have reported in this six months. This is, let's say something that we don't care too much but we do care what is the trend. We have more near-miss reporting on, let's say working conditions – on not having the proper equipment on board for doing the job safely, not, let's say spares or whatever reaching the vessel timely on a planned job, but they should be there, or the condition where it was, let's say ordered to do something despite the fact that the risk was still high. And this is what we are monitoring mainly. And based on these trends, we take some measures per case. OK for the next three or four, six months or even a year our focus is to minimise or get back into downstream line these things. Of course without [laughs] doing it but not reporting such situations. Report situations but let's avoid the reoccurrence of the situation.

Vasilis

We have a system in place for reporting of near-misses of accidents, incidents etc., etc. So we maintain a database of this and we collect them and we can see the trend which is reported to the vessels and in case we ... in case we let's say, see that we have the trend is getting up, OK, we respond to the captain, we are telling that something is going wrong here and we are trying to find measures in order to mitigate or stop this trend ... These accidents are analysed by the [Health, Safety and Environment (HSE)] department. Also we invite other persons from other departments for example if it was let's say, if an accident took place in the engine room we call the superintendent engineer, or if an accident occurred during discharging we'll call somebody from the operations department. Ok. And we identify the primary cause and the root cause of the accidents.

The approach differs from company to company but the essence of analyses is the same and is in all cases required by the audits and inspections/vettings of the safety management systems whether commercial or statutory. However, trend analysis, though mentioned by many of the interviewees, was still very much an accounting approach, what Woods and Cook call 'counting failure' (Woods and Cook, 2001) and not very focused on deeper level analysis nor sufficiently on precursor analysis in the anticipation of risk, a situation evidenced by the kinds of statistics presented by HSE staff to management. The full potential of information technology in the interpretation of data is yet to be exploited by the companies. This is necessary for as Dalan (2007, p. 1) puts it: 'outmoded paper-based systems, or the use of stand-alone data systems are helpful, but cannot achieve real results across the organization. Such systems often reduce safety managers to 'bookkeepers' rather than 'drivers' of fleet-wide improvements and knowledge-sharing'. In response to the observation of limited use of analysis of trends beyond descriptive statistics, Barry noted:

Barry

That is it. Analysing tools are not yet in place. No. Which we are in the process. We still have to modify some parameters on the reporting issues, because if you want to have analysis, free text is not a good thing. No. So we have to define the parameters which we want to analyse on and that's a job which I will be involved in to sit down and the … because today we found that there's too many. We're reporting too many and they are not specific enough. So that's a work which has to be done … That's why I say that trend analysis is on our programme, because risk events happening repeatedly, even if they are not severe, may be a final signal of a deeper problem.

Barry brings up the very important point of uniformity of classification as a prerequisite for risk data analysis: getting data that can be analysed appropriately and can be used for enhanced inter-organizational learning. Risk categories are at best company specific and in the extreme, case handler generated thus differing from case handler to case handler even in the same organization. The categories are also limited to accidents and near-miss events with little or no consideration of precursor analysis (as defined later in this work).

Evan expresses similar sentiments on behalf of Beta Gas Carriers regarding the limitation of usefulness to learning of just filing reports to show to auditors. He mentions SHIPNET which the companies in the Magna Group are introducing as a safety management tool.

Evan

And that [trend analysis] is the next step for us and that's what SHIPNET is designed to do – is analysis – and I absolutely agree that that's an area that we have to do better on because we're not doing it – we're not doing it [in an optimal way].

The SHIPNET system is already up and running in Alpha Tankers, but like in the case of STAR IPS in Olive Shipping it is not yet optimized for the kind of analysis that enhances proactive risk management.

Another part of the interpretation of information process is information selection. This is directly affected by the amount of information being received. Information overload – manifested in the multiplicity of checklists and procedural documentation – was evident in the companies.

Tim

I actually think you probably have, if anything too much information.

Information overload limits the ability of individuals to optimize the selection process and places stresses on individual and organizational risk control capacity and resources. In routine every day work individuals and organizations have to contend with significant amounts of information that all appear relevant and that create difficulties in the differentiation between 'real' signals and noise (Kiesler and Sproull, 1982; Marcus and Nichols, 1999). Organizations cannot commit resources to the investigation and assessment of every single incident at the same level.

One way in which an attempt is made to address this is to establish differentiation criteria based on perceptions of consensus notions of consequence severity. In the examples below, a colour-coded system and a numeric event classification system are used to distinguish between levels of risk information and to subsequently attract different levels of attention.

Barry

We have a risk matrix on board which is a three by three – a simple one – and the master when he puts events into the system (they have to) and it will end up in red, yellow or green. So let's say red, no green ones is closed on board. We only want at home for ... we want them reported for the awareness of the crew and we want them home to run statistics. The yellow ones will come in to a case handler here in the office which will be closed by the case handler and the red ones they trigger a professional team to go on board and do the deep root cause analysis which is not, let's say the seagoing personnel has not had the adequate training to do. But we have people here in the office which takes over that.

Tim

[If we had a spill for example] the first thing I would expect, in the system the first that I would expect to happen, you're going to have an investigation. If it is serious enough that you had a spill for example; we grade the level of whatever the occurrence from one through five and if it's anything above a level three – which a spill is above a level three; that's documented in our SMS [Safety Management System] – we have to have a shore based investigation as well as the onboard investigation. And then there will be an investigation report out of that and we would then determine if that was necessarily ... if it was serious enough for a safety bulletin to go out as well. So we now have a three stage process. We also have a monthly magazine which we send out so the basic stuff, the low level stuff that people have had a near-miss that we want to learn things from like, you know, somebody wasn't wearing his PPE properly when he was going to repair a crane or something like that – basic stuff that is (I'm using the word cautiously because it's not the right word to use but) 'routine' and it's non-dangerous, non-life-threatening. As I say these things are graded within our SMS – so there are clear cut boundaries. Those are routine stuff. We put those in our monthly magazines. Everybody can read and learn from. More serious stuff we would have the investigation, we would have the investigation report and if there were findings that were considered systemic failures we would update our SMS, we would update our procedures to prevent that failure from happening again and that's the key to the root cause analysis that we keep to.

Often the overload in data is not caused by diversity of risk information but a repetition of events. Some of the companies consider the quality of reporting especially of near-miss reporting as not being at the level where it indicates a fundamental understanding of risk and accident causation. It appears that the shore management are of the opinion, that for the same reasons that the seafarers (ship officers) are not involved in root cause analysis of the 'red' or 'level five' risks – being inadequately trained/educated – they are not 'able' to recognize more underlying root causes and report these.

Alexander

I mean I can tell you examples of near-miss reporting, but everybody reporting near-miss about personal protection equipment; somebody not wearing safety glass or safety helmet. OK it's a near-miss, but this is not a 'quality near-miss'. There are many cases where near-misses do occur but the people cannot recognize. Actually to me the near-misses, the way they are reported, it's a waste of time, because they are reporting the same and same things; PPEs and PPEs and things that individuals do … So near-miss, nowadays I wouldn't say that they are – at least the way they are reported and the quality of the reporting from shipping – I wouldn't say that makes any sense.

There was, however, a sentiment – at least in Metron Shipping – that the reporting quality was 'maturing' in relation to risk understanding.

Vasilis

No, no [we don't find that the reporting is repetitive and that we are getting the same things over and over again]. I think the reporting of the crew, the reporting of the crew nowadays is very mature than it was before and of course they respond very accurately, very carefully and without having in mind any blame culture. Ok. So there are free to report whatever is happening on board … In the beginning when the ISM was implemented, that was 1997, I think 98, whatever was happening on board it was reported as near-miss. Most of these reports were not supported, were not accurate, were not, were not anyhow for reporting. And so we were receiving about 1,000 let's say near-misses in the beginning – 500 or 800. Nowadays it has been reduced and the crew members are matured; they are reporting what they have to report. They are reporting what they are suffering, what it has been done to them. OK. Example, if they had, if they slipped on deck, if something was falling on his head, or if they have any behaviour problem, if they had fighting on board, if they had minor accident on board. Ok. Everything is reported on this way.

In both cases, the near-miss reporting appears to be predominantly reactive. The limitations of this dominance of reaction to historicity are discussed in Chapter 7 on near-miss reporting and precursor analyses.

Information Distribution and Interpretation

Computer-Based Versus Paper-Based

For the distribution of information there appears to be a progressive migration from paper-based reporting and feedback publications to computer-based risk communication modes. In many of the companies, the database system is also employed for all other production centres of the company, including purchase orders, maintenance and stock control. This helps them see risk not as an external appendage to the company's main operations, but as part and parcel of the efficient running of the ships. It makes it easier to bring risk issues into ongoing social discourse as a necessary parameter for productive company performance.

> Evan
>
> How do we disseminate that information? Again we, with modern communications and everything else today, we again have a system of fleet bulletins, where we would send out 'there's a particular risk' or problem that we want to identify to the ships. And those are also copied to our supervising office in Asia and they disseminate that information to the other LNG ship management companies within the Magna Group. If necessary, they'll send it to our safety group in Asia. And we also then send those to our training centres and make sure that the training centres and the instructors on board are current with the current problems, you know, that are happening on board so that when they have the people in for training, you know they're also dealing with relevant and updated problems that happen on ships.

> Paul
>
> The whole managed fleet use that system [computerized risk data management system in Olive Shipping] which means that some of the vessels in external management still use the system ... The officers will have access. They use it for various things. Could be even Chief Engineer looking at purchase orders for parts for the engine or pump man looking for something.

Edith

All the junior officers [also have access to the database]. Of course in each, this system they have modules and not all modules are accessible to everyone, like if I am using one system I can only access those things which concerns me of course, otherwise the security is no longer there. But if you are concerned in the particular incident or in the particular observation that was fed into the system your attention will be called by the superintendent and then you will be asked to answer that particular entry into STAR IPS. Normally here in the Philippines like because most of the crew onboard Olive Shipping ships are Filipino, as you may know, we have lots of, we have superintendents for a particular fleet of vessels, because the vessels, we have lots of ships and these are divided into fleets – we call it fleet – and then every fleet, there is a superintendent who is taking care of this STAR IPS, reviewing whether the entries has been answered and has been closed out and then they will call our attention. Like they will tell me that there is an entry in the STAR IPS that concerns your department. Please answer or provide me the answer and then I will create in myself. Something like that.

Evan

We have not implemented the SHIPNET what's called occurrence management. So we are. . But we will. It's in the plan for this year. We use SHIPNET for a lot of it, a lot of functions on board for example the planned maintenance system, machinery history system, the requisitioning system, the purchasing system, our accounting system. We've just installed what's called the performance management system using SHIPNET … We are actually putting into place now a new system for particularly incident reporting. Because what we find, what we are finding, is that you know, if an incident happens, everybody and their brother wants to, you know, to know about it and sometimes we've found that we haven't done as good a job in keeping all the various different stakeholders, you know, apprised of what's going on all through the incident and so we found that our procedures were not well enough documented; we found that our staff had not been trained well enough in the expectations of the customers and who wants to be informed and when and that's led us to review the policy and … that's underway right now, actually … So what we do with the near-misses is we, we report – I'm pretty sure – we report these to our supervising office in Tokyo. And they compile in a summary report.

Alexander

[We have a computerised] database which is open to all the vessels. Everybody has access, from the captain to the OS. We are administering the database but everybody on board the vessel has access to see what's going on … Well the file, the database is actually capturing the facts. Whatever is captured comes by email or attachment ... We also circulate them to the whole fleet on quarterly basis, apart from being available in the database.

Vasilis

Actually what we are doing now we're going to have let's say, ship–shore connection with this internet and with LAN let's say. We don't have for the time being. OK this is something that we expect to be done within the next two years, let's say. But what we are doing, we send them all the reports on a hard copy, we send them bulletin, safety bulletin, various other reports, accident investigation analyses ... all these we are sending them and they are implemented.

Formal Reports and Meetings

Formal reports and associated meetings are another way in which the companies try to disseminate information.

Paul

We have the PEC meetings; security meetings or standard meetings that they have on board, the whole crew to discuss the lessons learnt for the past that we have in our system that are new quarterly reports, all kinds of information that they have received on risk and risk management as well, new procedures. So when these lessons learned are published, for instance, they take that up in the first PEC meeting they have on board. For onboard safety awareness and risk management in that aspect, lessons learnt are probably the most important thing we have. So based on all the nonconformity, near-accident reports, accident reports, Jack or someone else will go through the reports; what happened, what do we need to be aware of, what do we need to focus on. So for instance, you have splash accidents for chemicals, get some splash in someone's eye or whatever, and then we focus a lot on safety equipment, safety equipment, this is important now! And they take that up in their meetings onboard.

Tim

In the office we have a monthly safety committee meeting. So items that are, items like that come up in the safety committee meetings. We also have the ship's, we review the ship's safety committee meetings and they may well present us with items that we aren't aware of or think need attention and they would go into high focus areas and the ships get copies of all of those. They can see what we are focusing on as well.

Evan

For example we always have the masters and chief engineers who come here for a full briefing, a full day's briefing, before they go on the vessel. So again we try to emphasise, listen, you know, we're here to support you, you know you're the guy that's out in the middle of the ocean, you know, in a harsh environment with a tough job and our job is to support you not to tell you every little thing to do.

Alexander

The captain, the chief engineer, the chief officer, the second engineer, the cargo engineer, some representatives from the crew depending on what is the risk that we are dealing with. So it's very, very clear for the vessel. In the office on the other hand, apart from the managing director who has the overall command, it is myself, the technical manager, the ships manager or the ships managers, if it is a broadly wider decision and it could be the crew manager, the IT manager, the HR manager or the media response manager. Contractors, not the manager. Sometimes we use also the claims and insurance manager. But these are let's say, on a case by case basis. Mainly it's the people who have [responsibilities for] risk. Ship manager, sorry, managing director, myself, technical manager, ship or ships managers and that's it ... Apart from the shore based arrangement, there is a master's committee, safety committee on board, where the vast majority of the crew participates, the ratings along with the officers, and then they can speak up officially and formally any concerns and issues.

Vasilis

Once we have an accident, the analysis of the accident the root cause and the lessons learnt to avoid recurrence are disseminated to the fleet under the shape of risk assessment, under the shape of safety bulletin. They are disclosed to them, what happened and once they receive something like that they call. Captain is calling a meeting – safety meeting and he demonstrates what we have given to them, what, how we proceed with the analysis of this accident and what lessons we learnt.

In addition to all the formal attempts that are made to capture and disseminate risk information, there are many instances of informal communication of risk information.

Professional Information Brokers

The shipping industry makes minimal use of professional information brokers in learning from risk. The only setting in which this is prominent is in the handling of crisis-related media exposure. Here many of the companies, in line perhaps with an intuitive appreciation of the social amplenuation of risk, make efforts to have consultants work with them in handling the media prior to, during or after a crisis. The media has become in itself a risk source which is incorporated into risk models used by the companies. They may also categorize the risk source from the media depending on geo-political factors and subsequently employ media-response contractors.

> Alexander
>
> Yes we had the case of two LNG carriers offshore United States but very close to the coast, and it was a big issue, also in the media and we did some, let's say exercise OK what we should pay extra attention to when approaching, let's say high media coverage areas. An incident in let's say, Nigeria won't have the same publicity as you have in Europe or in United States. This is a reality we have to face. So what extra measures do we have to have when calling in these, let's say high media coverage areas. And then we will put down some, again practical things, maintenance things, pertaining to any due procedures for our ideas.

Post-Distribution Data Interpretation

It must be noted that there is an interpretation process that goes on post-distribution. This also has the same dynamics as have been discussed earlier in pre-distribution interpretation. Seafarers receive the information that is distributed via the various organizational formal and informal channels and bring to this information their own interpretations based on subjective factors like their own perceptions, backgrounds and so on. For example in the case where a seafarer feels he/she is being sidelined for whatever reasons, the interpretations they bring to safety information/orders from superiors is tempered by the cognitive and affective associations they have with the superior.

> 43-year-old 2nd Officer – Male
>
> Sometimes I am ignored in regards to my opinion as a 2nd officer and a safety officer of the vessel because the captain ordered to do so. My duty as a safety officer is already been set aside.

> 30-year-old 2nd Officer – Male
>
> The 'Lead by Example' quality is missing from the senior lot. Often you may find a Master in shorts/flip-flops standing on the catwalk asking the crew why Safety Helmet is not being worn. It sounds like: 'we don't have to follow the law; we are the law'.

The post-distribution interpretation of risk information affects what value/merit the target subjects give to the information and the extent to which it solicits the subjects' behavioural association with what is intended by the information. In other words, though there is merit in the need for the wearing of safety helmets, this risk information is interpreted as an imposition of irresponsible control and is often reacted to with aversion and non-compliance.

Organizational Memory

Memory does not reside in the business or legal setup that may be seen to be an organization (nor its buildings), but in the human resource that makes up the organization and the organizational socio-technical systems that simulate processes of cognition and memory at any point in time. The dynamic nature of learning means that the potential for organizations to learn is irrevocably linked to the retention and turn-over rates of the human resource and to the presence of systems that facilitate storage (memory) and passing on of relevant information.

As part of the organizational learning process, the companies have various forms of system features that can be seen as representing organizational memory. Two of the primary ways put forward were storage of risk information and learning in manuals in the representative form of procedures, guidelines, rules and checklists. There is also storage in the form of documentation on previous accidents/near-misses in reports. The most pervasive media for storage were databases either computerized or paper-based.

Searchable Databases as Memory

　　Paul

　　Basically Star IPS is the system we have that the vessels have access to as well
　　as shore personnel. So among other things we have our procedures there. So all
　　the manuals are available from there and when we update these manuals here in
　　the office, we update them in that system and it will replicate overnight to the
　　database that they have onboard. And also likewise anything they do onboard,
　　the system will replicate over in the office.

Documentation as memory is severely restricted by human's capacity to locate context-relevant events that are documented 'somewhere in the system'. A searchable database is helpful in this regard, but the status quo with even the better systems, is one in which running optimum searches is still an evasive goal. The formats range from storage of risk related information in emails, word processor files, computerized databases and dynamic systems that are networked across all ships and shore offices. Most often the companies merge the use of all these.

Paul

Well, I mean OK you have the data sorting here. The problem with the system is probably mostly with searching. I mean you can find everything here but you have to know how to search. Like it is for every system I suppose.

This under-optimization of search functionality is partly due to a lack of clear event classification structures and language/semantic uniformity, even in the same company. The Olive Shipping STAR IPS system, for example, has a classification drop-down menu with various accident causation labels including 'human factors'. In the filing process, the association of this class or category with the event in question is at the discretion sometimes of the captain and sometimes of the case handler. The classes themselves have been generated by Olive Shipping and are not explicitly related to any research-based classification systems.

Paul

The captain will first or whoever reports it will first set the category on. The category might be changed by the case handler. The category [classification] list is a predetermined list. It is set by the company. Exactly who has done it I'm not quite sure. You can only select between these. You cannot write something else.

Tim

So we have the basic general information about the categorising, breaking it down. Was it personnel, was it environment or whatever? What happened? Where did it happen? What problem was it? Who on board investigated it? On what it was?

In most cases, the databases were restricted when dealing with very serious accidents. In 2004 the *Atlas Arrow*[12] belonging to Olive Shipping was involved in a major accident that made international headlines. This very serious accident is not to be found in the database as a narrative of what happened and lessons to be learnt as is the case with minor accidents.

Paul

Atlas Arrow isn't in the database anymore ... I think for the most part the reports that are sent out and taken care of and also the lessons learnt as well is the vessels' main source of information on what has happened before or so on, but for things like the *Atlas Arrow*, you know, the analyses, the investigations, the reports are so extensive and the action list required to implement to fix everything that you

---

12   The name of the ship has been changed.

can, has already been done within the system so, I think in a way it's already in the system.

Apart from searchable computerized databases the companies still resort to paper-manuals and 'information refreshment' by significant agents who have periodic contact with crew and shore staff for the purpose of refreshing organizational memory.

Barry

But all these accidents which is, we have HSSE [Health, Safety, Security and Environment] manager who is taking, and when we have conferences he is constantly going back to the big accidents that hit the company and for those officers we also will have it on the agenda when we are doing it because we have the training centre on Subic Bay in Manila. So all these accidents are on his agenda. So they are not forgotten. And they are taken regularly back into, for reminders. There is also always a safety bulletin taken off in reports. And we also have the – hopefully – lessons learned. This is going to all the ships. So I can read back to the, whatever. And we also have when it comes new lessons learned is added to this one. It's a demand that they are always taken up in the safety meetings on board. And they are always to stay in here for the crew to go through. So here is the history, if you want, just add-on as a history.

Paper-based memory is still very dominant in the companies with often negative consequences. Evan notes for example:

Evan

In Asia our colleagues are still very, very much in this culture of paper-based system. Everything is paper-based and they really struggle in their audits because they can't find it. Doesn't mean they're not doing it. It means they can't prove. And what I showed you – near-miss – that's the type of thing that the oil major man will say 'OK show me how your near-miss system works'; and they'll want to see. How do you make, how do you close this out? You know they'll pick one. And if we can't show that to them [then it suggests we are not doing it].

Given the volumes of risk-associated information being collected in these shipping companies, paper-based systems are severely restricted in the way they lend themselves to searching and also for analyses. Evan's description of the situation in their Asian offices is also indicative of one use to which 'organizational memory' is put, that is, that of referencing for the purpose of 'passing audits'. This remains a key focus of the industry as a whole and has the negative result of transforming safety managers into record-keepers – bookkeepers – rather than analysts. Paper-based memory also increases the administrative burden on ships and onshore.

23-year-old 3rd Officer – Male

> The company has its own considerations. They are too far from seafarers and from what is happening on board. Usually response to accidents is seen only in increasing number of formal paperwork and safety-related paperwork is formal of all. All we have is increasing paperwork which is a great risk itself. It is much harder that it seems and takes too much time. So we work non-stop without rest.

It does not seem realistic to expect that workers – especially those on board ship – will spend a lot of time perusing the many manuals all over the ship in what little time they have away from work. Even if this were to be the case, it would be grounds for the exacerbation of another kind of safety problem – that of fatigue and the associated dangers in contemporary shipping.

Procedures as Memory

In many cases the companies see the development of procedures as the most optimum way of facilitating memory in the belief that these procedures become the 'culture' of the company even where the particular incident(s) that generated the procedures are long forgotten by agents of the organization. The compilation of lessons learnt in databases exists as organizational memory only in the sense that it establishes reference points from which a safety culture may be derived. The companies operate with the mindset that the solutions – the lessons learnt – are incorporated into the operational culture and thus limit the need for continual reference to descriptive accounts of particular incidents.

Paul

> If you're new and you know the Olive Shipping systems – and you are required to know procedures – so I think you would automatically know these things [outcomes of lessons learnt] even if you did not know where it comes from [the specific incidents]. It will be, I mean procedures are revised on weekly basis and could be any small thing from a near accident, an accident that we've had, anything that could trigger the fact that we've got to revise the procedures.

Thus the accidents are synthesized for causative factors and subsequently mitigating measures put in place. This synthesis and the resulting avoiding or mitigating procedures, become the memory that individuals – as agents of the organization – draw from on a day-to-day basis. In the long-term this forms a culture that is accepted as dynamic and subject to new input and 'growth'. Again this dynamic and procedure-driven memory needs to have feedback cycles in operation and re-emphasizes the important role of communication in learning and risk construal and control. The procedural routines and rules thus become not just memory but 'carriers of organizational knowledge' and changes in them as 'processes of organizational learning' (Kieser and Koch, 2008, p. 330).

When Organizations Choose to Forget

In relation to organizational memory, an interesting observation was how the organizations handle very serious and extremely negative events that attract negative public attention. All the companies visited recognize, prima facie, the dangers of 'organizational amnesia'. Olive Shipping, however, has tried to quickly 'forget' the *Atlas Arrow* incident and all the very negative press it generated. In the company, the incident is not a major reference point and the assumption is that the organization has learnt valuable lessons from the incident and does not need to 'traumatize' itself with continuous and explicit reference to it. Importantly, what the company tries to 'remember' are the tacit and explicit procedural rules that were generated as part of learning from the event. This is in sharp contrast to the path taken by the Magna Group with respect to four very serious accidents that occurred in a single year – 2006. In this case the company management has tried to keep the memory of the particular incidents – not just the procedures that resulted from them – vivid in the minds of the employees. There is significant effort to do this with audio-visual media (DVDs, posters and so on) on the ships and in group offices. These media are regularly used for training and awareness-raising. Interestingly the benefits of such an approach is debated and is not entirely agreed to by key persons in safety management.

Tim

I have deliberately steered away from all that went on in 2006. And as I have, we have a, there's a safety focus programme on it, there's DVDs on it. I have deliberately steered away from it because I think it's now two and a half, three years on, why dwell on your problems? You were shaken out of your complacency. You've put in place systems that stop you ever being that complacent again. I think the systems are [helping to stop complacency]. I still see the ability to have complacency creeping in. I still see the attitudes that probably caused the, and I'm speaking purely from a lack of knowledge about what caused the original, and I don't want to know what caused the original deliberately. Because I deal with what I am seeing on a day to day basis, and that's more important to me. But I think that people like me, who are digging in, who are analysing, who are making assessments, who are not letting a single near-miss, not letting a single accident or inspection fly right by – that's my job – to stop those things from being missed, make sure that every single one of those things are investigated and find out what caused it and deal with it, and prevent that root cause and improve and share that with the rest of the group. So yes I think the systems that they have in place now, with people like me in the organization, the departments like mine – it's not me personally, but it's what I do, it's my task – will prevent the company from; Oh. I can't say will prevent the company from ever having a major accident again, because that's never going to be the case. But we will do our best to lower the likelihood of it. We will do our best to learn from these things and prevent them from reoccurring. I don't think you need to, I don't think you need to wallow in it. I have fears that we wallow in it. We have big posters on the ships – right on top of the gangway: 'Never forget 2006' and you've got a big circle that crossing it with four pictures of ships basically killing themselves, and I think that, I think it actually produces a very negative impression for anybody who's coming on board our ships – anybody who walks on board, a port state control inspector, a SIRE inspector … It shouldn't be anywhere posted around the ship. That kind of thing it just … yes, you needed to shake people out of their complacency. You needed to get people's attention. You needed to show that actually we had four major events in one year. And you needed to do something pretty drastic to get people's attention. But that was three years ago! … Pointless! Absolutely pointless! When you, when that poster is being on the bulkhead for more than a couple of months, (you can't even see it. It's part of the furniture). I mean people who see it are people who go onboard the ship, who've never been there before and they see that thing and they think oh my **** what the heck kind of ship am I sitting on. And that creates very negative impressions for people who are assessing us. And the people who are assessing us are these flag state people who demonstrate to the rest of the world that we have a poor performance or they are our customers.

Jason

With respect to all the [four major incidents], well I mean thankfully, I mean we weren't involved in any of them, thank goodness. But [the group head office] has taken you know the, I mean, been quite forceful in reinforcing these problems. I mean we've got posters and: 'remember 2006'. Like for every year since all these incidents they'll send out reminders to you 'don't forget what happened' and all these, like the scenarios that happened at the time. And we disseminate that information through our fleet as well so they know what went on, you know, what happened to these ships. Unfortunately we had an incident last year [laughs] so, but luckily, I mean the ship was damaged but there was no pollution or anything like that.

This author does not deem it necessary to declare which approach is better, only simply to note that proponents of each approach must bear in mind the merits of the other approach.

Additionally Tim's and Paul's comments coming from key individuals in safety management, is an example of how individual perceptions and voices, even of leaders – may be lost in mainline organizational policy-setting. The reasons for this may include the fact that the individuals in question possibly did not consider the issue critical to safety or the possible existence of other dominant voices (as expressions of power and control). Leaders also have the need to voice out their perceptions – whether laterally to their peers, upward to higher organizational echelons or lower to subordinates (Ashford, Sutcliffe and Christianson, 2009) and these 'speaking up' and 'speaking out' dynamics are also influenced by team psychological safety.[13]

Transactive Memory

The notion of transactive memory makes 'group training' as relevant as 'individual training'. Wegner (1986, p. 186) describes 'transactive memory' as 'a set of individual memory systems in combination with the communication that takes place between individuals'. Kieser and Koch (2008, p. 334) – citing Liang et al. (1995), Wegner (1995) and Wegner et al. (1991) – note that:

> According to the concept of transactive memory, groups, with the help of directory knowledge [defined earlier on the page as a 'directory of people who may have knowledge relevant for changing certain rules'], can simulate a common memory and thus achieve a higher degree of memorizing capacity than members acting individually.

Human limitation is such that, it is not possible for any individual to retain optimum knowledge and memory of all contributory systems and factors that pertain to any given risk situation. Where teams are trained together they develop a knowledge of knowledge positions or a cognition of cognition centres or meta-memories. This means that team and organizational members draw from a collective memory which is dispersed strategically within different individuals or units. Training teams together therefore 'provides workers with a valuable resource – a transactive memory system – that facilitates knowledge distribution and coordination within the group' (Liang, Moreland and Argote, 1995, p. 386). In practical operations, this resorting to transactive memory is constantly done in shipping with the division of labour (and the associated memory) on a bridge for example, or between bridge and engine room or shore units and shipboard units. With training for transactive memory, however, the shipping industry, for the most part, finds it impracticable to intensively train crew members for example in important operational scenarios that can only be simulated off the ships. To some extent this challenge is mitigated in the context of ship crew by the regular holding of drills. Significantly transactive learning/memory opportunities between ship and shore are few and far between.

Edith

> We [train crew] individually because you cannot ask the whole team to come down and have training, because in Olive Shipping we don't allow like if chief

---

13   See Section 7.6.1 on team psychological safety and leader inclusiveness.

officer is signing off, second mate or third mate cannot sign off together with chief mate. There shouldn't be any new, no new sign-on crew – two new sign-on crew in one department at the same time. So there should be, what you call it, the period that has to be observed before another mate sign off. So you cannot send them together, all together. The training is individual. Whenever they are on vacation and they have a skills gap identified through the performance evaluation or through their own willingness to attend then they will be sent to training.

Tim

Our philosophy on ship teams as a whole is we don't have fixed teams. People don't go back to the same ship on a routine basis for a fixed period of time so training ship teams ashore as a whole team would never happen under that philosophy.

Personal opinion, yes [there are merits to training people together as a team]! Most definitely! Personal opinion I don't think the model we work here is best model. I've come from a Ro-Ro [Short Sea] shipping background and a Ro-Ro shipping background you keep the same teams together, you put same teams on the same ship for as long as you can. They do one [month] on, one [month] off and they go on leave together and they come back together. Yes. They're stuck on the same routes all the time so they know the ships, they know the routes and that helps you. I personally prefer a hybrid model of the two. One of the things we found with short sea shipping is when people have been on the same ship six, seven, eight years the challenge is gone. It is not as good as it could be. Performance drops. People get complacent; they get comfortable. They don't push themselves as much as they could. I actually prefer, what I think would be the best would be a hybrid model. When you have ship teams who come together for two or three years and then you disband that team and they go elsewhere so that you mix them up and they form another ship team with different people on different ships and then you get the best of both worlds. You get the influx of people coming in to change ideas, to bring new ideas, to improve things, but you also get stability of the team coming together and performing together and knowing how they perform together. And I think with bridge team management for example which the safety and protection in bridge team operation is human factor, is people saying is this the right thing? Did you mean port instead of starboard? … Those ones work better if you have a team. That having been said, provided you have good trainers and you train the same training model and the same training course for every team that goes through the bridge team training course, it doesn't matter. If everybody does the same training with the same people, doing the same thing, it doesn't matter! Because you've trained them the same, as long as it's implemented the same on the ships it's not a problem and that's where you step in with your shipboard inspections, with your auditing routines to make sure that it is implemented. So there are merits to both processes. There are merits to training teams together and I personally think it's better to do it that way. But in economic terms, it's not better to do it that way. In economic terms it's a huge waste of resources. To pull an entire crew, to pull 20 people together at any one time to do training on basic stuff that they can learn with anybody is a huge waste of resources; huge waste of their time as well. Because you're basically, the only way you can do that is to grab people together when they are all on leave. The chances of your ship team being all on leave at the same time – zero. From a management perspective, I don't want my entire crew changing over all at once. There's zero corporate memory and that's dangerous. So the actual, the idea of pulling a whole crew together to do training all at once – never going to happen. Not sensible. There're too many problems with it.

Alexander

[Training ship crew together, like in aviation where you take one ship crew off at a time] is not practical. It will be ideal. Ideal! However, not the same crew sign off the same vessel – people with the same people from the same vessel. I mean you can't take 20 people out of the vessel and then put another 20 people on board. It's totally discontinuation and you will have an accident – definitely. Then it's people living in Philippines, Houston, Greece, Ukraine and again all these people when they've disembarked they have pre-scheduled commitments. Somebody's wife is delivering a baby; somebody is disembarking because of the unfortunate passing away of a loved one. So ideally it would be perfect. If we could, but practically it cannot be done. On the aircraft it's very limited; they're still located on the same place. For shipping, it doesn't work. Even if it's not the money – the question of money; not at all (for this company at least) – it cannot be done logistically. You cannot do it. Plus practically you cannot take 20 people out of the vessel and then put another crew on board – trained crew, competent crew. This is a totally separate vessel, even if it is a sister vessel. The little things that an engineer knows about this machine the other engineer doesn't know it. He will find them out but [at what cost]? And that's why SOLAS calls for additional training if 25 per cent of the crew change at any port [SOLAS Chapter III regulation 19 paragraph 3.2 (IMO2004)].

It is difficult to engender significant team training off ships in a bid to increase transactive memory in a particular team, for the reasons that have been coherently pointed out by respondents. In general personnel turnover is a problem for transactive learning or memory (Liang et al., 1995, p. 392). As indicated by the comments of the interviewees the extant nature of ship crewing practices and operations offer more challenges to transactive learning in shipping. The shipping solution has been to seek to design shore-based courses in as uniform a manner as possible and to ensure that specific persons in certain roles have the same knowledge and perspectives. This solution is reinforced by the long-term evolution of hierarchical ranks with associated competencies and specific roles. Though there are significant merits to the calls for changes in the organizational structure on board ships (see for example Barnett, Stevenson and Lang, 2005), this would be one more hurdle to overcome – the speedy inculcation into the global maritime culture and the prompt awareness of the roles, knowledge and competencies of any new structural roles. Another solution – taken seriously by the companies in this research but possibly missing in many other shipping companies – is a high focus on human resource retention which tends to increase members' familiarity with other members' roles and memories in the specific company.

Change and Adaptation

Feedback loops should be seen to be resulting in change, where necessary, and not be left at the level of espoused theory as is often the case:

> 23-year-old 3rd Officer – Male
>
> There are many formal ways of such learning. Formally [in theory] it works but actually does not. All we have is increasing paperwork which is a great risk itself. It is much harder that it seems and takes too much time. So we work non-stop without rest. But actually nobody pay attention on the real things. They think that paper is enough to provide safety. Example we have bad safety shoes. They are slippery. Company was informed but answer was that shoes were approved fine [emphasis added]. So they would not change. Formally shoes are safe: really, not. All the other things the same.

The ultimate goal of learning in the context under study is the optimization of practice. The companies visited are concerned about practical change resulting from their latent or explicit learning. The most significant change in the learning processes come from new or updated procedures that are supposed to inform action.

> Barry
>
> We have the explosion and sinking of *Atlas Arrow*. Total loss. That's one [thing that led to change], not barely nine years ago which affect cleaning procedures immediately ... And we earlier also had explosion on another one but all of this – especially within the specialised tanker industry has changed our way to handle things as fires and explosions. And also you have the way of treating slops today. It has been a process of learning and adjusting and improvement over the whole period since I started and up till today.

Some of the change has been in personnel for the explicit purpose of bringing in new safety/risk perspectives. In the case of the Magna Group, learning from a major accident led to significant organizational structural change.

Jason

I mean I'll be honest with you it's only like since Dima [new manager] joined us about 18 months ago – October 2007 he joined – and he brought a lot of fresh ideas and there must have been a lot of resistance to this in the office as well because he's quite forward thinking and the company was becoming very staid. It was very I would say conservative with a small c and it sort of, he has brought a lot of fresh ideas, kicked a few backsides [laughs] you know, got people motivated more, you know and I think it's helped a lot in the last 18 months. A lot of people have gained a lot.

Tim

In terms of learning I think they, before I joined the company, 2006 was a disastrous year [regarding the incidents on four ships in the Magna Group] ... I think they were shaken to their very core. I think that before the incidents they were very complacent. I think they felt they knew what they were doing and they were very good at it and they didn't change, didn't want to find anything to improve.

Yes [my employment was the result of this]. Absolutely! Absolutely! This is the kicker. They did not have – as a company – health, safety and quality department till then. We did not exist as people. We were considered unnecessary.

In the tanker-owning companies, to ensure the effectiveness of change, formal change management systems with required feedback loops have been incorporated in the safety management systems. This is a requirement of TMSA (OCIMF2008).

Evan

We obviously have a management of change ... and this is something that we try to strongly encourage, you know, the ship staff to, you know, to use this management of change, so if they find something that, if they find something that in the manuals that is not correct, a procedure that's been written, something that's been asked to them by a charterer or somebody else that they don't feel comfortable with [they can bring it up]. One of the most important pieces of this management of change is the feedback loop. In other words you have to, before you can close a management of change, you have to gauge the effectiveness of the change. The other thing that's required here is that the people who the change is impacting, you have to give them a chance to be part of the process.

> So we have to go to the people on board the vessel and say 'look this is a change we're thinking about making. Are you OK with this? Do you see something that we don't see?' So we're required under the procedure to capture those ... We would probably [consult the master and chief engineer] that's who, the senior managers on the vessels. I mean what one hopes is that if this involved a change to mooring procedure, whatever that they would get the bosun involved, they would get, you know the second mate and third mate who are actually on the bow and stern and collect their comments and feedback.

Often the change may not be explicitly practical but may be one of attitudes. Alexander discusses the aftermath of a major incident that had somehow escaped the safety system and safety managers

Alexander

> It was two months discussions and committees and questions how such a major thing could have slipped our attention or the captain not realizing that he should have taken permission from this office. I wouldn't say that major changes occurred out of this; however what happened actually is that, we tried to spend more valuable time on briefing and debriefing rather than generalities and routines – how it was your vessels, your food, [etc.]. If there's anything they will tell us. We try now in the briefings and debriefings to focus on quality issues.

From the foregoing it can be observed that change in the companies is ongoing. It may be the result of learning at a very routine level or may be a result of a significant variation. In most cases it is the result of specific incidents, but most of the companies try to afford opportunities to company employees at all levels to request change as per their perceptions. In the safety realm, feedback from the latter is not as common as that from specific incidents especially when the change involves the commitment of significant resources.

Chapter Summary

The processes of knowledge acquisition, information interpretation and distribution, organizational memory and the resulting change that are discussed in this chapter are inherent in the companies' safety management systems. They are not explicitly labelled as such but are part of processes of risk identification, assessment and control and are espoused to be inherent to ideals of continuous improvement.

The companies exhibit congenital learning in how they draw from philosophies and traditions tied in to their inception. Experiential learning remains the most dominant mode of learning. It is evidenced in knowledge acquisition processes that focus on reports from the ships. These reports are significantly biased towards reaction to incidents (near-misses or accidents) with very little precursor information (as defined in this work). Even for near-misses, not only is volume of reports an issue, but quality of reporting is questionable.

Vicarious learning is not as pervasive or systematized as experiential learning. In the former, third parties, particularly related to commercial interests, play a dominant role. Reliance on IMO and national databases for vicarious learning is relatively not pronounced.

Distribution of information is mainly via computer and paper-based reporting, databases and meetings. Distribution is enhanced by ship visits by high level management staff, but this is not as systematic as other forms although for the tanker industry (those subscribing to TMSA) it is part of the requirements. It is noteworthy that risk data is subject to subjective interpretations pre and post distribution. There was no evidence of in-depth interpretation of data by way of analyses beyond the counting of incidents and descriptive statistics. This limitation in the analysis approach was acknowledged by some companies. Such companies accompany this awareness with plans to increase the versatility of their data systems, to inquire into more underlying patterns of risk progression and accident causation specific to their contexts.

Organizational memory is mainly characterized by procedures which purport to capture lessons learnt and integrate them into organizational culture. Other memory forms are information held in various databases and in individual and department competencies. The memory held in databases was found to be severely restricted in the extent to which searching was possible as a recall mechanism.

Finally the learning, no matter how limited, does result in some change at some point or another. There were a lot of espoused theories of safety though and some of the comments from survey respondents indicate that these sometimes do not match with theories-in-action.

Chapter 7
Discussion of Research Findings: Emergent Themes (I)

A technological society has two choices. First it can wait until catastrophic failures expose systemic deficiencies, distortion and self-deceptions. Secondly, a culture can provide social checks and balances to correct for systemic distortion prior to catastrophic failures.

Rajendra Pachauri[1]

Purpose and Outline

The risk-associated learning processes discussed in Chapter 6 are impacted on significantly by a number of themes that emerged from the research. These were labelled as emergent themes and include near-miss reporting vis-à-vis precursor analysis, communication, accident causation philosophy, crew engagement and welfare, team dynamics, role of culture, the notion of entropy and drift/migration to the safety border, and human resource retention. In this chapter the emergent themes are used as points of departure to discuss the findings of the research.

Near-Miss Reporting and Precursor Analysis

The research shows the primary focus that the shipping industry places on near-miss reporting. The most evident benefit of near-miss reporting is the opportunity to examine and learn from accident causation dynamics without the attendant costs in lives, property or environmental damage associated with accidents (Budworth, 1996).

Heinrich's (1931) often-cited accident pyramid (with ratios indicating the occurrence of 300 incidents with no injuries, to every 29 with minor injuries and 1 major accident or lost time case) is often used to emphasize the opportunities offered by near-misses, as is the adaptation of that pyramid based on Bird and Loftus' research on 1,753,498 reported accidents (Bird Jr. and Loftus, 1976, pp. 33–4). This latter research suggests that for every one serious or disabling accident there are 10 minor injuries, 30 property damage accidents and 600 incidents with no visible injury or damage (near-misses). As put by Bird Jr. and Loftus:

---

1   In a speech (as Chairman of the Intergovernmental Panel on Climate Change (IPCC)) to the United Nations General Assembly on 24 September 2007.

The 1–10–30–600 relationships in the ratio would seem to indicate quite clearly how foolish it is to direct our total effort at the relatively few events terminating in serious or disabling injury when there are 630 property damage or no-loss incidents (near-misses) that provide a much larger basis for more effective control of total accident losses. The valuable loss control potential that exists in the information available to organizations that expand their investigative effects to include ... near-miss accidents or incidents with no visible injury or damage is substantial. (Bird Jr. and Loftus, 1976, pp. 34–5)

While the ratio may not be replicated exactly in the shipping industry, the general import is deemed to be applicable to that particular context (Kuo, 2007, p. 8). Despite these benefits of near-miss reporting, it is nevertheless the case that the industry is arguably not maximizing the learning potential from them.

Evan

We're not capturing all the benefits of the near-misses. I mean BP; I went to a safety conference with BP last year ... It was very interesting, you know, what they said was, you know every time there is a major incident, we of course we put together a team to investigate and we come up with the root causes and, you know we spend a lot of time, effort, money for these major incidents. But what we weren't doing was putting the same type of resources against serious near-misses. We weren't investigating them. We were doing what you saw – OK somebody looks at it, yeah, OK, that's good, bang, you know, let's carry on. And so BP actually put together a team of four or five people, took them out of their operations in engineering disciplines and safety disciplines, moved them over into the HSE department. And that's all they do: is investigate accidents, but more importantly near-misses.

While stating this, Evan also acknowledged that BetaGas Carriers was not yet committing substantial resources to the investigation of near-misses, describing it as the 'next step' for the company.

Despite the benefits that are to be gained from near-miss reporting, there are limitations to focusing almost exclusively on near-misses and accidents which seems to be the predominant case- driven mainly by legislative instruments and external pressure from commercial interests. Proactive risk management should optimally be dependent on precursor analysis rather than solely near-miss reporting. In some texts the word 'precursor' is used synonymously with 'near-miss' or 'near-accidents'. Corcoran cites the US National Academy of Engineering, for example, as defining the term 'precursor' as 'any event or group of events that must occur for an accident to occur in a given scenario' (Corcoran, 2004, p. 80), and himself defines it as 'a situation that has some, but not all of the ingredients of a more undesirable situation'. Similarly Carroll defines the term as 'events that must occur for an accident to happen in a given scenario, but that has not resulted in an accident so far' (Carroll, 2004, p. 127). Using such definitions, they then include near-misses in the category of precursors as indeed does the whole publication both references are from. The view taken here – and germane to the contribution of this work – is that of precursors, not necessarily as specific events, but generic and antecedent conditions of accident causation. In this view a 'near-miss' is a 'quasi-accident' and not a precursor. The *Oxford Compact Dictionary* defines precursor as 'that which runs or goes before ... precedes and heralds the approach of another' (Murray et al., 1991). The two – that which heralds and that which is heralded – are not the same. Near-misses are accidents in essence with only stochastic variations leading to different outcomes. A precursor is not dependent on chance. It is an undesirable but stable condition that has the potential of contributing to the kinds of dynamics that lead to near-misses and accidents. For example, a lack of conditions that allow for giving feedback in a communication system, a lack of worker engagement, general issues of fatigue, communication difficulties via language use and other such 'conditions' are precursors of accidents, but not in themselves, near-misses as events. Precursors herald both accidents AND near-misses and do not rely for their differentiation from accidents on outcomes.

It was agreed by all that the value of a near-miss is in its replication of the dynamics of accident causation, but without the attendant loss. It is therefore perceived as a golden opportunity to learn at least cost: what Exxon has called 'free lessons' (Reason, 1997, p. 119).

A pervasive view in the industry is expressed by one respondent as follows:

A respondent

> Near-misses are not accidents. It's just that too many near-misses might lead [emphasis added] to an accident OK. That's why as I told you before, we maintain statistics and every six month we tell them, we are telling them to the crew, to the vessels, trends. We have near-miss in terms of personal protective equipment, in terms of navigation, in terms of management on board, in terms of wear and tear.

In the view of this author, the value of near-misses is real and the actions described in the quote above are very laudable. However, the premise for the actions described is flawed. The idea that it is a multiplicity of near-misses that lead to an accident seems to emanate from perhaps a wrong interpretation of Heinrich's and/or Bird and Loftus' work. It is debatable whether Heinrich or Bird and Loftus intended any such connotation of causation of near-misses *leading* to accidents. It is also not clear whether the ratios relate to accidents and near-misses that are similar in causation dynamics or are just numeric quantifications of reported happenings. It is noteworthy that Bird and Loftus caution that 'it should be remembered that [their research] represents accidents and incidents reported and not the total number of accidents or incidents that actually occurred' (p. 34) and that Heinrich's work has come under substantial criticism (see for example Manuele, 2003, p. 122 ff.).

Near-misses **are** valuable and should be treated just as seriously as accidents are whether or not they exist in the exact empiric ratios noted by Heinrich and Bird and Loftus. However, any set of ratios should be seen as being predictive only in the sense of **potential** and *not* in any temporal sense. Complacency in the presence of an inordinate number of near-misses based on the perception that there will be many of such *before* there is a serious accident is flawed at best and outright dangerous at worst.

It is in reference to this predictiveness – in the sense of potential for accident causation – that it is suggested that near-miss reporting is limited in its value to true proactive safety management. Responding to near-misses is just as reactive as responding to accidents and as has been pointed out many accidents happen for the first time without the presence of earlier near-misses or other accidents (March et al., 1991; Sagan, 1993; Taleb, 2007). The focus of risk anticipation should be proactive engagement with precursor analysis. To use an example adapted from Budworth (1996), imagine a scenario where a brick may fall from a scaffold to the ground below. A number of different outcomes are possible. The text in brackets indicates how reporting is qualified in the majority of cases:

1. The brick does not fall. No one notices the danger it poses in its position.
2. The brick does not fall, but someone notices the fact that its location increases the possibility of it falling – the danger it poses.
3. The brick falls, there is no one in its way, and subsequently no one notices anything out of place.
4. The brick falls, there is no one in its way but someone notices the next day that it has fallen and could have injured a worker (near-miss).
5. The brick falls and narrowly misses a worker below, who is not wearing a safety helmet (a near-miss).
6. The brick falls and narrowly misses a worker below who is wearing a safety helmet (a near-miss).
7. The brick falls and hits a worker below but (s)he is not injured because (s)he is wearing a safety helmet (a minor accident – relatively).
8. The brick falls and hits a worker below who is injured because (s)he is not wearing a safety helmet (a major accident – relatively).
9. The brick falls and seriously injures a worker below despite the fact that (s)he is wearing a safety helmet (a major accident – relatively).

At what point in this sequence of increasing severity of consequence should safety management seek to engage itself? What kind of reports have to be made if safety actions are to be proactive? What kind of investigative processes, mitigation choices and associated training would be subscribed to? The example can be extrapolated to ones involving more substantial losses in personal, property or environmental terms. Arguably proactive reporting and safety action are better served by a focus on the conditions in scenarios one and two prior to even a near-miss (as they are reported) occurring. Relying unduly on near-misses is unnecessarily extravagant and opportunistic – perhaps dangerously so. It must be stressed that near-miss reporting has its place in risk management and all the arguments of increased reporting and the creation of the environment for reporting and a safety culture are valid. However, it is also important to check and address what is being reported and what is not being reported even in the best of safety cultures. Even if the system supports the unhindered reporting of historic incidents (near-misses), it still would not address the functional system features of normal work and the little deviations that are normalized over time and that are only glaringly obvious in hindsight. The predictiveness that is required for proactive risk management would only come about when those 'little deviations' and their normalization are recognized as precursors by individuals who think in that paradigm and where the functional system allows for, accommodates and uses such reporting. If this argument is plausible and has merits, then it points to the necessity of enhanced training and education that make it possible for operational and management staff to recognize underlying conditions – precursors – that create potential for accidents and very importantly the commitment of resources to this. Furthermore the kind of reporting that is (or will be) required by a focus on precursor analyses demands a high level of team psychological safety. Necessarily, such reporting is about situations and contexts that are subjective because they are a priori assertions of conditions that have not resulted in dangerous events or undesirable outcomes – yet. Unless there is an appreciation of such non-tangible or immeasurable phenomena, there will be no credence given to such reports.

In the preceding account of fictional scenarios, existing incident reporting in shipping focuses on the kind of scenarios in four, five and six and accident reporting on seven, eight and nine. Scenarios five and six are the ones most reported (see an example of such a report in Figure 7.1).

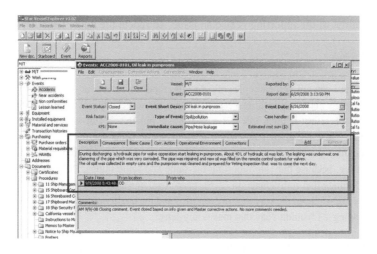

Figure 7.1   Incident report showing primary-level reactive action

The figure shows how an incident report is closed without any checking of the obvious potential for underlying factors. The fact that the statement 'proudly' states that the pumproom was cleaned and ready for vetting shows the role that the tanker inspection and vetting regimes are playing – rightly or wrongly – in motivating for safety (see Chapter 8 on resource availability and utilization, organizational entropy and drift).

There are other issues to practical near-miss reporting that the research found. Some companies require a certain number of near-miss reports to be made for a specified period.

> 23-year-old 3rd Officer – Male
>
> Example is near-miss reporting. If something actually happens to be reported this way, better not to report as the only reaction will be blaming somebody. But there is a requirement (unwritten) to send three near-miss reports per month. It is evident that such occurrences not necessarily happen that often.

In the opinion of this author, this is a completely incorrect approach to reporting. Tim agrees and states:

> Tim
>
> From a safety perspective, I am dead set against that [mandating a number of near-misses to be reported per period]. I do not believe you get honest data. I have met companies which have said 'ah we require five near-misses from the ship per month and two of them must come from junior officers'. And you get them because you told the ship that it's a requirement and the data is meaningless.

The 'quality' of the reporting is also an issue. To repeat the quote from Alexander:

> Alexander
>
> I mean I can tell you examples of near-miss reporting, but everybody reporting near-miss about personal protection equipment; somebody not wearing safety glass or safety helmet. OK it's a near-miss, but this is not a 'quality near-miss'. There are many cases where near-misses do occur but the people cannot recognise. Actually to me the near-misses, the way they are reported, it's a waste of time, because they are reporting the same and same things; PPEs and PPEs and things that individuals do … So near-miss, nowadays I wouldn't say that they are – at least the way they are reported and the quality of the reporting from shipping – I wouldn't say that makes any sense.

And Vasilis in describing even the 'matured' reporting (as compared to the early days of ISM implementation) in Metron Shipping had this to say:

Vasilis

Nowadays ... the crew members are matured; they are reporting what they have to report. They are reporting what they are suffering, what it has been done to them. OK. Example, if they had, if they slipped on deck, if something was falling on his head, or if they have any behaviour problem, if they had fighting on board, if they had minor accident on board. OK. Everything is reported on this way.

Reporting what they suffer is quite a way away from proactive risk reporting.

Double-loop and Single-loop learning

The kind of learning that is associated with near-miss and accident reporting appears to be predominantly one of single-loop learning as shown in Figure 7.2.

It further appears that this is the reason why without significant near-miss or accident-inspired legislation, the industry maintains a state of lethargy premised on tradition and how things have always been done. It is surprising that an organization as substantial as the Magna Group is – among the top 10 carriers of the world and established in 1917 – had up until the occurrence of significant accidents in 2006 no set-apart department for safety and health, and that it took those accidents to incorporate such expertise into the functioning of the organization. The lethargy is also noticeable in that there were shortcomings that were pointed out by the managers interviewed – such as the aforementioned limitations in near-miss reporting and the under-utilization of trend analysis potential in information technology – which though acknowledged as important, are not necessarily a priority. This is perhaps because the issues at stake have not been prioritized by the imposition of external legislation or commercial pressures.

This kind of mindset can consign the industry to a Sisyphean cycle of compliance to new bursts of legislation and frantic efforts to comply whenever there is a 'new' accident or global shift in commercial pressures.

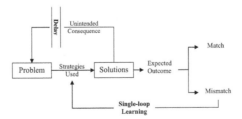

Figure 7.2   Problem solution model
Source: Kim, D.H. (1993)

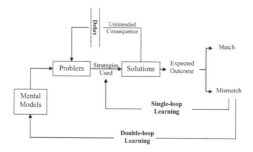

Figure 7.3   Problem articulation model
Source: Kim, D.H. (1993)

Theoretically the desired learning approach is as shown in Figure 7.3 – a combination of single-loop learning with double-loop inquiry. The single-loop paradigm is basically a reactive one and is evidenced in the companies by the substantial focus on near-miss reporting.

There are some reasons why the latter double-loop approach is not that obvious in shipping. Primary among these is the fact that current officer (and shore-management) training does not appear to address the fundamental inquiry mindset that questions mental models. Despite the contemporary proliferation of university degrees attached to the long-standing competence qualifications, maritime education and training is still very much one that focuses on teaching individuals 'how to do things' and not necessarily augmented with 'why things are done' a particular way. Competence is still defined by ability to repeat what has always been done, cognitively and behaviourally: the honing of particular skills without questioning of underlying paradigms. This is valuable, but given the constant repetition of the same accidents, one would surmise that a different (augmenting) approach is called for. The industry recognizes that there is a higher level of risk awareness that is not gained from MET (Maritime Education and Training) today.

Arguably, many contemporary quality standard systems in their present form do not substantially lead to continuous improvement (an essential requirement of a learning organization) in the sense of marked innovation. They rather refine and entrench old routines – classic single-loop learning – with the potential result of stagnation and resistance to positive change. While this is still learning and may be better than deterioration, it is not up to the demands of a complex, multicultural setting where there may be rather rapid change in the external environment (whether technological, regulatory or societal). Optimization of performance, greater anticipation of risk scenarios (Woods and Cook, 2001, p. 93) and negotiation of reality (Berthoin Antal and Friedman, 2005) are better afforded by double-loop learning. To achieve holistic double-loop learning, the seafarer must be in the loop and his/her questioning of fundamental premises of operation encouraged and solicited. Satisfactory acknowledgement of such questioning and appropriate answers and follow-up actions lead to identification and engagement. In this way organizational performance in risk management will be enhanced with ship/shore personnel appreciating better the dynamics of organizational learning and with organizations working to gain positive outcomes from contemplated and intentionally designed learning processes: deutero-learning.

It is noted, however, that the problem may not be so much the inability of the industry to learn in a double-loop fashion as the priorities that are established in light of resource constraints and commercial pressures and profit-making. This issue of resource availability and utilization is discussed in Chapter 8.

The Critical Role of Communication

As indicated in the preceding chapter, the shipping companies in the research do learn (to different degrees) along the lines of the major themes denoted by Huber.

Primary to this is *communication*. It has been noted that the kind of information that is communicated, when and to whom it is communicated are often a choice prerogative that lies with top management (Child and Heavens, 2001, p. 313). This may be true in most cases where information is accessible to top management. However, when it comes to risk, there are significant (even critical) pieces of information that may rest with lower organizational levels. The role of leadership (meaning here top management) is the leading of the creation of an organizational culture, supported by systems and procedures, which will facilitate uninhibited disclosure of risk-relevant information, while filtering out irrelevant information. As Child and Heavens subsequently point out (2001, p. 314) 'the challenge of leadership in facilitating organizational learning therefore lies in maintaining a judicious combination of both control, in the form of guidance and resources, and the autonomy required to motivate knowledge-generators and encourage the free flow of information'.

The systems that have been put in place by the companies purport to optimize the communication links in the company to enhance the transfer of risk information. However, it was clear from the research that the communication hinges on what management considers appropriate even apart from the necessary filtering out of irrelevant information. Despite the espousal of measures for upward communication, existing communication is predominantly one-way in nature. That is not to say that there is no ship to shore communication, but that the bias is in the shore–ship direction. As one seafarer put it:

> First of all, communication between company and seafarers isn't done frequently. That is big problem. Company always give us only email. This isn't 'communication'. This is only 'notice'. (23-year-old apprentice officer – Male)

Additionally, the quality of the communication from ship to shore is predominantly one of 'reporting' and not the interrogation of the status quo. Ship–shore communication is also 'filtered' through the master.

30-year-old 2nd Officer – Male

> The Seafarer-Company relationship is more like an interface and less of a nexus as the seafarer is directly in touch with the Company officials during sign-on/sign-off. Onboard a ship, it is always a passive relation as all the relevant communication is routed through Master (notwithstanding the screening of messages and the discretion of transmitting the message verbatim).

The reasons for this are that the Master is perceived as just that – the *Master* – the expert who commands and controls the ship and is accountable for this control. Be that as it may, as Dörner (1996, p. 169) notes, 'experts see things in much more differentiated form – that's what makes them experts – and for that very reason they may overlook other perspectives'.

The communication of risk information is thus mediated by the social stations that comprise individuals with exposure/non-exposure to diverse risk information that the seafarer is ignorant/knowledgeable of. In a way what this implies is that a reliance on experience is stressed to inordinate proportions and contributes further to a hierarchical separation that is based more on *rank* than *nationality*. Nationality differentiation was not seen to be as significant an issue as rank differentiation in the research. Indeed when asked what factor was in their opinion highest on a company's priority list when credence had to be given to different risk perceptions, the seafarers ranked *rank* as the most significant factor. *Time spent in the company* was ranked second and *nationality* a distant sixth. Though the quantitative results showed means of OECD officers' team psychological safety to be statistically higher than those from the rest of the world, the effect sizes were not such as to overly emphasize the role played by nationality. It is accordingly felt that the effect of team psychological safety in global shipping today is associated more with hierarchy of ranks than with national background.

Semantic differences in language between ship and shore may also constitute a difficulty with risk communication. What this implies is that the shipping industry always benefits from the employment ashore of people who have had sea experience and can relate to the 'language' of seafarers and see things in a manner more tuned in to a 'sea perspective'. Evan emphasizes this at length:

Evan

> When we say no blame, we mean no blame. And that's why I repeatedly say, you know seafarers are really no ***** type of people and they'll look at some letter and, but what they're really looking for, what they're really looking for is what are you doing. Not what you're saying, they're looking for what you actually do

when it happens. I've had a, quite a long discussion with our commercial people here and in Asia about the authority of the master and, you know what I say to them is 'look, every time you question the master's judgement, OK, particularly when it's a commercial matter and you're unhappy about the costs of that, every time you do that OK you erode his confidence OK that he has the authority to make a difficult decision. OK and if you do that repeatedly, over and over and over again, I can tell you what's going to happen. He won't make a decision. There will be some very dark night when things are really going to hell OK and he won't make a decision. He will want to talk to you about it before he makes a decision and that can lead to a disaster'. And I feel like the owner of that problem and I am constantly on the look out for messages from our charterers or our commercial people, you know to our masters that I feel undermine. He's made the decision and we should support that decision. If we want him to change that decision, it means that we have to give him more information. Why does somebody make a decision? If we believe that he's the right person for the job – and hopefully we've hired the correct people – then we have to support that decision. If we feel the decision should be different, it's most likely because we have different information than he does. OK. We're in the office. We have access to resources that a guy on the ship doesn't have. And so if we want to have him change his decision we need to give him more information so he can say 'oh, I didn't know that. You're right. Maybe we don't need to do this or maybe we can go on'. Maybe we can get in touch with the maker of that pump or machine or valve or whatever. We can say OK we checked with the maker and they say it's OK to operate in manual or something you know, you know like that. And I very, very strongly believe in that. And I see it, you know, I was a master for a long time and I always felt like our company supported me even sometimes I made very difficult decisions, but I always felt that they would back me up and support me. And so from where I sit here I always try to do that as well and I always want the masters to know that. And that's one of the messages that I give them every time they come through the office, is, you know, if you're not sure, if you're uncomfortable, you know, we've got a, you know, and this is again, it's important to have the company staff the office with people of appropriate experience that have been there so we can, that's how we can support the vessels. OK. Because we've been there. And this is a problem with some companies because they don't want to pay, to hire ex-mariners and have them ashore, you know requires a certain level of, you know level of compensation, you know to get, people interested in that. So if you're just going to hire, you know a guy who, OK he can sit there in that chair but that doesn't necessarily mean that he can do the job. Some cultures, you know, think they can do the job. I always like to say, the most dangerous people are the people that don't know what they don't know. And they're going around and they're undercutting and undermining the people on board the ship, you know, with stupid and unhelpful comments. Anyway I see that as my main job [laughs].

Another seafarer adds:

> It is company's people [ashore] that do not have knowledge of on board or experienced. So some people don't understand what crew say. Another people ignore saying of crew (22-year-old A/O – Male).

The statements above show the mediating role played by social station and even geographical location in the sharing of risk information and a suggestion as to how the potential negative consequences can be mitigated – by the employment of people with similar backgrounds as seafarers. Another suggested solution is the temporary employment of ship staff in offices.

Both solutions attempt to create an appreciation of differing perceptions of risk and to ensure that risk communication is not limited by ill-understood views, language and inaccessible islands of information. While these measures are worthwhile and definitely beneficial, it must be cautioned that this should not evolve into groupthink as discussed in Chapter 2. The necessity of retention of different perspectives is critical to safety. What is important is the ability to communicate in a manner unrestricted in scope or language, but which allows for the expression of points of view which may even at the extreme be considered disruptive. It is when organizations learn how to assimilate such views without a disruption of the risk communication fabric, that they can gain access to information that may be critical in saving the company from serious danger. The skill of negotiation of reality is pertinent to organizational learning (Berthoin Antal and Friedman, 2005). Some of the companies try to engender an informal approach to information sharing, whether related to risk or crew welfare. With due cognizance of nationality differences, they try to do this across cultures with different meetings all round the world. This is done by the companies visited but it doesn't appear that it is a general practice in the shipping industry.

One determinant of communication is theorized to be **culture**. This was one of the most emphasized factors vis-à-vis learning and risk construal.

Cultural Interaction

Quantitative results of the research indicate that ship officers do not think that nationality is an issue when it comes to the credence given to risk views by the companies. Secondly, experience of multinational crew did not significantly predict change in team psychological safety. It is clear, however, that there is recognition in the companies of the role of culture (often perceived as adverse) in mediating risk perceptions and learning. Nationality causes other team-affecting social dynamics, but does not seem to significantly impact team psychological safety. The companies try to mitigate the consequences of such perceived adverse effects of differences in culture by primarily training, efforts at human resource retention and cross-cultural and shore-based crew interaction. Some of the respondents felt that 'training' was the most effective way to deal with cultural differences but even where such views were expressed such training was either non-existent or minimal. One of the approaches to solving these cross-cultural difficulties has been the use of training courses with basically share lists of stereotypical items that are deemed to characterize particular cultures. This has been criticized in the literature and recommendations made as to the use of negotiating skills rather than superficially learning stereotypes. Jason acknowledged this and narrated a situation that perhaps describes how many shipping companies view culture and the training needed in that context:

Jason

No [we do not have training in communication for the seafarers]. Not really. I mean, I wouldn't say we haven't but I mean nothing pronounced. We have been doing like inter-cultural training here in the office [for office staff] but we haven't, I mean, we should, we probably will have to do something because I remember years ago, when I was with [earlier place of employment named], we had like a two day course. It's quite strange actually because the first morning was like understanding the Filipino [laughs loudly] and the next day and a half was understanding the Japanese [laughs long].

As humorous as Jason found this, it seems this is still a pervasive approach to cultural awareness training in the industry in general. Legislatively speaking, the most relevant IMO course addressing this issue is the PSSR – Personal Safety and Social Responsibility – course which is mandatory for all seafarers as per regulation VI/1 of the STCW (Standards of Training, Certification and Watchkeeping for Seafarers) Convention and Table VI/1–4 of the STCW (Seafarers' Training, Certification and Watchkeeping) Code (IMO2001). The model course is recommended to cover 14 hours (IMO2000). Many companies (and educational institutions) seem to treat it just as the example Jason gives – a stereotypical shopping list of dos and don'ts for particular cultures. While this may be suitable for tourist visits to a country for a few days, it is for a number of reasons, not suitable for people who share shipboard life for months on end and particularly for ship officers. The only other reference to such subject matters is in Table A-II/2 of the STCW Code which requires management level officers to have a competence in personnel management, organization and training on board ship and the related international and national regulations. Due to the complexity of culture as discussed in Chapter 3, and the associations it has with organizational learning, it will be beneficial when the topic is addressed (possibly within the requirements of the STCW) at a more in-depth level.

Culture (at a macro-level) is not supremely dominant in human relationships. Andrea (speaking about the relations between North-West Europeans and Filipinos on Olive Shipping ships) further notes the difference that individual personality makes.

Andrea

> I think it is very individual what sort of person the Norwegian is or the North West European is. If he were a very social person like I have some of the officers who have, are in a learning position and they've said oh they've learned heaps from this Filipino officer that was the second officer or he taught me lots of things that I didn't really know. I've had lots of benefit from being on this boat; I don't want to be moved because he is here. So they have intercultural learning aspects there as well.

Considering the complexity of culture (it is not possible to – cognitively – get to know all about any particular culture even where one predetermines who one's ship mates will be) and the uniqueness of personality, one solution would be to increase team and organizational level communication – as suggested by Berthoin-Antal and Friedman (2005) by training and education in the negotiating of reality: skills that allow for the openness, discourse and interaction necessary for efficiency in a high-risk industry (Osland and Bird, 2000). Better training will not only better intercultural communication but also inter-rank communication.

Quite apart from the difficulties of intercultural communication, another perceived difficulty was differential treatment based on nationality. Sometimes this was a perception on the part of the nationalities (as in the Andrea quote below) and other incidents were actually organizational policy (as for example the statement by an officer regarding salaries).

Andrea

> I agree that there must be team-building between the ship and the shore and I feel also … the culture difference like we have a lot of Filipinos and we were just at the Philippines – couple of weeks ago – and we had a talk to the cadets there that were going to school and some of the things – one of the things that I thought they were wonderful being so open with this – was that they feel that they wanted for us to take back to Olive Shipping and try to do something with this like attitude for the Filipino cadets because they feel that the Norwegian cadets or the British cadets they are taken more seriously. They learn the job that they're supposed to do while the Filipino cadets felt that some of the officers treated them like they would do the dirty work but not learn … The culture there as well and I know that Filipinos, they talk among themselves before going to anyone and telling them about their problem especially the officers or to the captain. So there is also a communication problem there. So I have understood from several officers that the Filipinos, they maybe found out about something the back way that's happening inside the Filipino crew; maybe some person that is going together with the other person and there is conflict, which may cause also problems on board with other issues and they don't find out about this

directly because they don't get in touch with them. So there's also the cultural side of things.

40-year-old Chief Officer – Male

Why do shipping company's pay foreign nationals especially whites more than their local counterparts even if both have same international ticket and take cognizance of his advice than the blacks?

While there may be organizational justification for this – possibly because of national manning laws – the practice remains and is perceived as indicative of racism by some. This substantially reduces engagement and as the research shows would then compromise organizational learning. The expectancy and equity theories of motivation – which augment organizational support theory – give substantial support for this. One respondent, for example, after responding quite positively to learning in his organization had this to say:

The answers above may be reverse while working with officer who strongly believes in racism and also believes officers must be white.

Due to the sensitive nature of this, it is seldom addressed as a race issue. The observation, however, still remains that this is a matter that continues to compromise communication, learning and hence safety.

All the companies consider high retention levels as key to breaking down cultural barriers and indeed for the tanker industry, TMSA checks this as a key performance indicator. As will be discussed later, this not only augments smooth intercultural interaction, but the organizational measures applied to increase retention, also serves to boost worker engagement. High retention rates help to engender an organizational acculturation process that limits the role of national/ personality differences. The organizations, by getting crew and shore staff together in informal settings, by clear and constant communication of organizational values and goals, and by individuals' exposure in experience and time to the organization's existing culture, try to bridge any cultural differences. Metron follows this religiously and has over the years developed a multigenerational workforce.

Vasilis

Metron has very strong relations with the officers and crew members OK. That is why we have so high retention rate. OK because we have, I mean we have very good relations with the seamen. They enjoy our, let's say benefits through all these years. We have two generations on board the vessels – especially from Philippines side. Fathers who have their sons who became now captains and we're looking after them in case somebody is sick or in case somebody he needs a loan or if there's any family problem, direct involvement and we are trying to mitigate the problems.

Difficulties remain, however.

Barry

Yes. The [organizational] culture is there ... I would say that's Olive Shipping's culture. But I would say that the ratings the Filipinos with us and most of the junior officers, but still the Filipinos have their own culture to carry and they are not at a level of openness as we wish them to be today. They still have some way to go and that's our job as a company to introduce. But I would think that if we are looking into a global seafaring culture, I think we are at the upper end of this safety culture because the crews we have on board our vessels have been mostly with us since early eighties. But the full openness which we want them to be is not yet there.

There is still a clear 'us and them' mentality, understandable considering that the managers and those trying to limit the influence of the cultural differences, see through, and are themselves influenced by, their own subjective cultural 'goggles'.

Paradoxically, the 'us and them' difficulty is also observable in a rather interesting setting. In one company, the officers appear to have so identified with the company that they had in some cases tried to do things that 'protected' the company from 'external bodies' like Port State Control authorities.

Alexander

I tell you when I came to this company years ago, it was like a club and by the years you feel like a family member. Now this thing is giving us, is giving people sometimes to do things that they believe they are for the good of the company, but actually they not for the good of the company. That's why it's clear there. They did because they believe it's good for the company, no. We want you to do your duties as we want you to do it. Don't think of the good of the company. We will take care of this.

The companies also try to limit the perceived negative effects of cultural diversity by restricting their manning needs to only a few nationalities.

Vasilis

We have in dry cargo vessels we have full Filipino crew. And we have all in tankers the majority are Greeks. We have Bulgarian officers and Filipino crew and we have five vessels with Filipino crew ... OK we don't have multinational crew members [implying unrestricted nationality] because in case ... because in such a case there are communication problems, some cultural problems etc., etc. We have only three nationalities right now: we have the Bulgarians, we have the Greeks, we have the Filipinos. You know the Filipinos are with us for many years since backwards 1967. More or less we have the same mentality. Bulgarians are with us since 1993, so they know each other. We have high retention rates so we don't face religious or cultural or language problems. OK the common language on board is English language and we don't have such problems in communication or in understanding each other or in mentality etc., etc.

Because of the desire to inculcate organizational culture into the crew, these companies are eager to engage officers at the lower ranks.

Alexander

First of all if it's a cadet [who is employed], we love that, because he will be growing up in our philosophy. If he's already a captain but for some reason we need to find out on the market, then of course we see all his cv, companies that he has served with, we make some phone calls, but if we see that he has been with companies that – of course we know, what's going on – and they know what's going on in this [area] at least in Piraeus. If we know that he has been with companies that have safety means to them just presenting themselves, but underneath is totally nothing, then we ensure that if we need definitely – and all the other qualifications are at the level – then we ensure that this captain goes with the best available safety-focused chief officer and the best available safety-focused chief engineer, to push him, let me say by their every day friction; and we monitor. one, two, three services if he is improving or reaching, or at least dealing with our needs OK he will stay in the company. If he's coming from a company that safety to them is a non-spoken word, then there is no space. There are cases where good people, they need to work and they go to this company, because they need to support their families.

This 'coaching not poaching' approach is at odds with a pervasive approach of the 'poaching' of senior officers by companies which do not want to invest in training especially in light of the perceived shortage of officers ('Bishop blasts poaching culture,' 2007).

The danger with a successful organizational acculturation process is that of groupthink as discussed in Chapter 2. Members of communities that are tightly knit in cognitive and affective terms may not be disposed to tolerating divergent views. As has previously been discussed, this is dangerous to safety and should be mitigated by the inclusion of team psychological safety as defined in this work and the overt encouragement of the airing of divergent views together with acknowledgement and explanation of what is done with these views as part of the organizational culture.

Team Dynamics

All the issues raised, in relation to communication and culture, are particularly pertinent in teamwork. The social dynamics that influence teamwork are critical in the achievement of productive learning. Team psychological safety and leader inclusiveness are two such relevant social issues.

Team Psychological Safety and Leader Inclusiveness

The following is an account of how reports sometimes get to the offices. It shows the '*in cognito*' measures taken by crew sometimes to get the shore office informed about what they perceive to be risky.

> Jason
>
> It's quite funny [laughs]. We had an incident. It was nothing much to be an incident ... A photograph appeared. Actually there was, actually this is an incident of a whistle blower. Actually I just remembered it. It was, we received a photograph ... about 18 months ago, something like that. And what it was a picture appeared from an anonymous hotmail address of two ... I think it was a Captain and a Chief Mate or Chief Engineer – I can't remember now – standing on the bridge wing, behind a barbeque [laughs] and you know there was nothing else. It was brought to the attention of the two people involved. It was a bit difficult to see whether it had been faked or not. But no, it hadn't. We found out it hadn't. In actual fact, I mean the people that were involved we were quite prepared to, you know keep them on but they chose because of the embarrassment and everything to leave, you know. I mean they were going to get like a reprimand because it's in our safety management systems: you're not supposed to do barbeques ... on tankers. I mean the ship at the time was gas-free and empty, and so they sort of. . They would obviously have to be reprimanded for this because it was a breach of our safety management system.

Despite stories like this and perhaps due to the effects of a lot of Bridge and Engine Room Resource Management courses and the increased effects of globalization, the quantitative findings show a relatively high mean for overall team psychological safety of 3.95 on a scale of one to five. However, for the one who sent this picture (and possible for the average ship officer), it would have been difficult to directly confront the ship master in the case above, especially where they believed the master must have appreciated the inappropriateness of his action and given the nature of onboard organizational structures.

The structure has its origins in pre-classical times. Meijer, for example, writes about the period before 500 BC when Greek sailors on Triremes had a significantly hierarchical organizational structure in place:

Apart from the 170 oarsmen there were another 30 crew members, including five officers, the *trierarchos*, the *kubernetes*, the *proreus*, the *keleustes* and the *pentekontarchos*. The *kubernetes* (helmsman) was *de facto* commander of the ship, since the captain (*trierarchos*) was a political figure who seldom came on board. The *proreus* supervised the front of the ship and kept a lookout for possible enemies and submerged rocks. The *keleustes* supervised the rowers, their training and feeding. To encourage the rowers in a fighting situation the *keleustes* was assisted by a *trieraules* who piped the time on a flute. The lowest in rank was the *pentekontarchos* ... who had all sorts of odd jobs. The other 25 crew members included a carpenter, marines and archers. (Meijer, 1986, p. 39)

The hierarchy is still very entrenched in contemporary shipping:

Andrea

I think from my experience in the shipping industry the hierarchy has been here for so many years, so the captain as the head of the ship and all the way down and it is a very respected hierarchy, and maybe it has to be that way as well for the ships to also be run efficiently and the communication lines are supposed to be as clear as they are. But from my experience working in other places the hierarchy is more a flat system where it is not as 'right up' as it is in the shipping industry maybe.

Barry

Still a shipboard organization is not a democracy, you know. It is still a hierarchy. So [communication can] be a problem yes. We are aware of that.

Jason

I mean from my point of view, I think if you got a situation on a ship, everybody knows their place on the ship in their rank structure I suppose but there should be the feeling amongst everybody that if they think something is not quite right they can always go to somebody higher up or even to the Captain to say that 'Captain, I've got a funny feeling about this. Should we not go this way or something?' He will be the one that makes the decision, but everybody below him should be able to approach him to give their suggestions. So that will be the ideal situation I think.

Tim

It is a hierarchical management system that we have on the ships. I don't think it's a problem. I know people have for a long time considered changing the way we manage ships and trying to bring different management styles to ships.

But actually what we need them to do hasn't really changed and the level of experience that people have is determined by the rank that they're given by certificates of competency. So basically, the right people in the right job is what we do. And I don't necessarily think we need to go around reinventing the wheel. I just think we need to actually train people a little bit better [laughs]. You putting in a new management style ... it's not solving the problem. It's just changing the way you look at it. And that is window dressing and not solving your issues ... Possibly [a flatter hierarchy will be more attractive to the youth], but I'm not sure it will serve the industry. We often adapt our systems to the circumstances we are under anyway. So, for example, on some of our ships we don't employ third mate, we only employ two second mates, because we want people of more experience in those positions because they're doing work that requires that much more experience so when there is a commercial need or an industrial need to change the management team it does happen. We did same with engineers. To be brutally honest engineering in terms of a hierarchical management really on modern ships is irrelevant. You have a chief engineer who manages the team and then you have two maybe three engineers who are not watchkeepers – they're day workers; they're UMS [Unmanned Machinery Space] workers and they are responsible for certain pieces of equipment. Now whether the first assistant engineer looks after the main engines and the 2nd engineer looks after the generators or the other way around in reality is pretty much irrelevant. It's just the task that needs to be done. So the industry has, by the nature of developments, by the nature of change, has already changed the hierarchical management system. We just keep the ranks; we keep the names. And in reality in the engineering terms a set of generators is probably, needs, might be less experienced in knowledge than looking after a main engine. So again it's done by competence, it's done by experience and generally what the industry has evolved into is a management team that actually does what it needs to do. I have never ... I have sat down and thought about at some point whether a flatter management style would work for the industry and maybe just because of the fact that I grew up with it, maybe because I'm used to it, but I don't necessarily think we will benefit from all the change. The other point is a nice flat management team structure isn't realistic ... We don't do the same in companies ashore. We have managers, we have teams, we have people in management.

Vasilis

I think the captain is the proper person on board [to control affairs on the ship and communicate with the company]; we've got to have a hierarchy on board otherwise everybody is going to be the same; we are going to have a . No. Ship is something different. It is not like ashore. Somebody has to take the final decisions and somebody has to make this final report, somebody has to investigate what happened. OK. So it is not possible two officers take decisions. Then we will lose the control. This is our view.

As desirable as a less hierarchical structure may be, the views indicated above suggest that it will take quite some effort to substitute the current structure given its antecedence in history. The current structure, however, does encourage some dissonance in communication. The research found that team psychological safety was correlated to leader inclusiveness ($r = 0.599$; p. $< .001$) and leader inclusiveness correlated to worker engagement ($r = 0.682$; p. $< .001$). The three were all significant predictors of organizational learning. Similar to findings in the medical field (Nembhard and Edmondson, 2006) it was also found that the higher the rank, the greater was the team psychological safety, the difference being statistically significant at the p. $< .001$ level. Contrary to popular anecdotal evidence, team psychological safety differences based on nationality differences (between OECD and non-OECD respondents) had a lower effect size (0.27) than team psychological safety differences based on rank differences (between management and operational level ship officers) with effect size 0.38. Furthermore, the differences based on nationality was significant at p. $= .003$, while those for rank was significant at p. $< .001$. This suggests that the strict hierarchy in shipping today constitutes more of a challenge for increasing team psychological safety than do nationality differences.

Given the entrenched hierarchical arrangement and its purported negative influence on team psychological safety, the question can be raised as to what would be a potential approach to mitigating these negative effects. Arguably what is more easily achieved is an enhanced view of the existing hierarchy and the team dynamics and communication patterns within it. This new view could be informed by more 'leader inclusiveness' as defined and discussed in this work. Ranking should not be a barrier to communication, but a way in which the cycle of optimum knowledge is transferred by way of experience, coaching and mentoring in an open manner. The best approach to this would, in the opinion of this author, lie more with appropriate long-term education about teamwork and leadership. By long-term, one means much more than two-week courses.

To help create environments where speaking up is encouraged and easily achieved the companies try to get crews together in informal settings.

Evan

The Magna Group holds a safety conference every year – an annual safety conference. In fact one's coming up next week, over in Eastern Europe and they hold one in the Philippines, they hold one in Mumbai in India and they hold one in Europe and I read through the minutes of all that and believe me you know, in Europe, the European seafarers have no problem speaking their mind even to the top management, they will get right up there and say well, you know. But when you look at the ones from the Philippines and India ... very few comments or if there are they're quite positive, you know … here is a, probably, a fear about speaking up, you know.

Perhaps what generates this limitation in speaking up in some cultures is the long-term association of different nationalities with different ranks and not the nationality per se. Edith notes how some junior Filipino officers were even more hesitant to speak up with senior Filipino officers, actually preferring to work with OECD senior officers.

Edith

Yes in Olive Shipping we're really encouraging the open and honest communication. I mean everyone is allowed to speak up. Regardless of who you are talking to. So I guess that has given them confidence to speak up. And another thing is that you may find it that like weird but some of the crew they prefer to have North Western European Captains. Some of the Filipino crew they don't like full Filipino vessels … because sometimes you will line them up to a full Filipino vessel they say 'no can I have a vessel with a Norwegian Captain' and then I will ask them 'why?' 'Yeah because I think it is easier to deal with Norwegians because they are because they don't have this like, what do you call it, the pet system. Filipino I think they tend – we tend because I am a Filipina so I should say we – we tend to have this some kind of pet that if we like this guy and then we, we like this guy. Something like that. It is normal I think for a Filipino. It is common knowledge that Filipinos are like that and we are trying as much as possible to answer that one.

Where ship officers are silent, it appears the quietness is one of quiescence as described by Brinsfield et al., citing Pinder and Harlos.

Silence is a form of communication that in itself involves a range of cognitions, emotions, and intentions. Additionally … employee silence has different meanings depending on whether its underlying motive is quiescence or acquiescence. *Quiescent silence*, for example, refers to an aversive, conscious state in which individuals purposefully withhold concerns, information, or opinions about organizational issues, even if they are aware that strategies other than silence could improve the situation. Despite their awareness of alternative options, these individuals are unwilling to take action because they fear that the consequences of speaking up will be personally unpleasant. On the other hand *acquiescent silence* refers to submissive acceptance of organizational circumstances and a reduced awareness that alternatives to silence exist. Thus, whereas quiescent silence is a state typified by high cognitive dissonance and apprehension [resulting for example, in a crew member sending an anonymous picture to the office], acquiescent silence is characterized by indifference and hopelessness. (Brinsfield, Edwards and Greenberg, 2009, p. 20)

One of the possible reasons for ship officer quiescent silence in shipboard teams is probably that the hierarchy on board is a competency-based one. Subordinates feel that questioning superior's actions on the basis of one's own perceptions is a direct questioning of the superior's competence.

Andrea

I feel that some people they do have, may see that things could have been done differently but feel like this has been this way for so many years, who am I to come here and tell them what to do in a different way.

Sometimes this perception persists even where the superior is open to such interventions.

Andrea

But in the one case this week there was one man who said that he felt that things could maybe have been done differently, but he was reluctant to talk about it. But I've got in touch with the captain and said that this accident has to be reported and reported to the right authorities and they're going to do that next time they're both on board. So he has ... the Captain's attitude was that he had no problems with it. He could fill out all these forms and everything. But it was the feeling of the person that was really the nature of the problem because I don't think really he would have had a problem had he taken it up with the captain. But he was sitting with that feeling, 'no I don't want to come in with criticism'.

One solution to this may lie in incorporating such TPS-associated inquiry as an index of competence and inculcating this value in an educational context. Over time what this will do is to raise issues about one's competence when one does not critically appraise relevant actions.

While teamwork considerations are given some attention when they relate to shipboard teams, the recognition of the ship–shore interaction as teamwork is not that pronounced. In many cases the sentiments expressed below are only too typical.

The company is NOT interested in hearing and learning anything from us seafarers. It's us against them, two teams that really don't work well together.

Anything we contact them about, we get negative response back. So we choose to have as little contact with head office as possible (36-year-old Electrical Engineer).

This statement reflects quiescent silence, not because of issues of competency, but issues of access and perceptions of organizational support. Where companies do not go out of their way to ensure that such access and communication lines exist, are clear, and are known, this kind of silence quickly becomes a culture that new entrants assimilate.

In the literature, the issues of team psychological safety are often expressed in terms of subordinates' voice and contribution to the team's work (Edmondson, 1999). Hypotheses are framed in terms of how the leader can contribute to alleviating the factors that limit subordinate contribution as for example leader inclusiveness (Nembhard and Edmondson, 2006). Rarely is this addressed in terms of the leader's own psychological safety. Leaders too are affected by similar dynamics in their relations with peers, their own 'leaders' as well as their subordinates. In a number of cases, interview respondents in this research (who are undeniably leaders) were themselves hesitant in the expressions of their views, and even under conditions of assured anonymity, were 'choosing their words carefully'. They were even more hesitant to raise some issues with other leaders although they made it clear that their views on those issues differed from company policy. This is in keeping with Ashford et al.'s view when they note that leaders have to decide 'carefully when to constructively challenge the organization … what new issues they introduce into the organization and when to introduce them. Lateral voice involves careful choice about framing, language, and timing issues' (2009, p. 197).

The Blame/Accountability Debate

Humans are not perfect. The systems they create are not perfect and to exacerbate things further, excellent closed systems, working together with other excellent closed systems in an imperfect world with open systems, generate complexities that create opportunities for safety breaches and even disasters. There are merits to not residing perpetually in communities of 'whodunit' but to merge islands of information and see operators, managers, legislators, administrators as stakeholders in a quest for safety. An ancient African saying[2] remonstrating against a 'blame culture' declares that *'deē ōkō nsuo na ōbō hina'*: to wit, 'it is the one who carries the clay pot to get water who is most likely to break it (and not by the armchair critic at home)'.

The increased focus on organizational factors rather than 'human factors' for accountability for accident causation appears not to be an effective dissolution of an unjustified blame culture but a reframing of who to blame and the disguising of this in new language. Blame continues to be dominant in the attenuation of risk signals (Susarla, 2003). In the shipping industry, together with the developing trend of criminalization, it serves as a very effective block to the communication that is required for proactive risk awareness and control. Hindsight is always 20/20 vision. When it is not explicitly acknowledged that the people who are involved in compromised safety situations, may not necessarily set out to get into those

---

2   The language is Ashanti Twi from Ghana.

positions and that those courses of action or inaction that lead to accidents, are justifiable rational choices in context, then ship officers and others like them will always be (perhaps justifiably) cautious about what, how and when they report.

23-year-old 3rd Officer – Male

Company has its own considerations. They are too far from seafarers and from what is happening on board. Usually response to accidents is seen only in increasing number of formal paperwork and safety-related paperwork is formal of all. Example is near-miss reporting. If something actually happens to be reported this way, better not to report as the only reaction will be blaming somebody. But there is a requirement (unwritten) to send three near-miss reports per month. It is evident that such occurrences not necessarily happen that often.

43-year-old 2nd Officer – Male

Of course the company learns from us (seafarers) because without these seafarers, the shipping company is nothing. But sometimes the seafarers are always to be blamed in case there are some accidents on board a ship. In this kind of situation accident already happen but the only thing the seafarer need is their compassion.

The way the company views blame and accountability is partly dependent on what accident causation philosophy it holds – tacitly and explicitly. If the organization and significant agents in it fundamentally understand accident causation in terms of 'human error' rather than 'system conflicts and non-resilience', they will have a greater tendency to blame operational staff. Explicitly, most of the companies subscribe to a 'no-blame culture'. However, in many cases there remains a difficulty in striking a balance between blame and the accountability and responsibility that goes with professional competence in a high-risk industry.

Jason

I think we are, I would call us a pretty fair sort of company. We have a no-blame culture and … I don't think we're fully there yet, but because as I said to you before there might still be a bit of unwillingness on the some part of the junior officers or the crew to come forward but I think you know, we try to be as open as possible and if they do have that problem I mean they can just pick up the phone.

Vasilis

First it is one of our standards or it's our position not to have blame culture. That's why we have people – we have crewmembers who OK, who suffer some accidents or incidents, either in terms of hardware or in terms of software resulted of a loss. OK what we are doing is we call him here, having analysed the accident first. OK we call him here, we talk to him on a smooth way – on a relaxed way – and we try to show him or to persuade him not to do it again, because the company suffered this and this and this and we would like to know his opinion why it happened. For example a violation of separation zones; a breakdown of a hydraulic motor despite the fact that we have given specific instructions how to sail within separation zone, despite of giving instructions how to drop the anchors. Despite all these verbal instructions, despite all these communications prior and after why it happened? OK this is not – we are not blaming him but we've got to know his intention. We tell him OK; 'we have in place this, this, this, why it happened? What was the reason? We want to know'. And of course we give him chances, but if the accidents are due to them, then we've got to take measures. Otherwise finally we will be blamed by the industry. This is something we would not like to risk, because first of all it's company's reputation. We maintain a very good reputation amongst our clients and we've got to maintain it. If one time we have one problem; the other time we have next problem or other problem then in these cases we start losing face. OK. And I mean we exercise hard efforts in order to reach this point so we don't want to ... Of course we don't blame our, I mean, senior officers, captains, chief engineers but of course we have communicated with them in order to ask why it happened, despite that we ... instructions were in place.

Chrestos LNG takes what in their view is a realistic position:

Alexander

First of all we have gone away from the 'no blame culture'. To us no blame culture means totally irresponsibility. We want people to be accountable. As we are accountable for our actions and omissions, they are also accountable. If you are an assistant cook or the captain, it doesn't matter. You have to follow certain procedures. You are aware of the procedures. You have signed that you have read the procedures and understand them. You have the right to ask for more clarifications or training if you are missing something, but we don't take that I did not do that, because I didn't know it. I didn't know it, while you are certified and have proven that you know it on your previous services, is not an excuse. Now for reporting an unsafe behaviour or practice, there is no such a blame culture to the reporter. But to the people that is under question, why they made this actions, yes there is no blame culture. If they are accountable and they have done something that they shouldn't have done it, of course we take into consideration some other factors; why they have done it, what is the motive? Personal gain? They believe they do it for the good of the company? They knew they are doing something bad, but they do it on purpose to help the company? It depends on what is the motive and based on the motive – we have a very nice flow chart – I can show you – based on the motive, then we have clear actions. Discipline of course is the last action where somebody do it, do it on purpose to harm the company or because he was lazy for personal gain. This is clear discipline, but if it was a one-time error with no serious consequences or because of some other, let's say, situational things, we may call in for additional training or coaching. OK disembark from his vessel, bring him to the office, talk to him, explain him that his behaviour or action is not what we expect from him and give him another chance, because after all we are all human. But if it falls under, let's say, the factor under consequences, even it was from good will, but the consequences of his actions or omissions will be leading to a serious accident on a scale – scale accident or incident – then he'll be dismissed. I will show you the flow chart, how we determine the actions, basis, individual. I believe this flowchart also is not on blame, I mean let's say on accountability. It's also on make people ... example ... I mean do something good. Again the same flowchart goes. How quote unquote this good action, exceptional action of the individual and then it goes up.

The flowchart mentioned is shown in Figure 7.4.

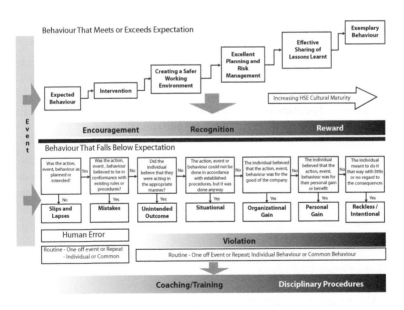

Figure 7.4   Enhancing safe behaviour model: Blame/accountability dichotomy resolution in Chrestos LNG

In most contexts the practical outworking of a no-blame culture vis-à-vis accountability from competent officers is fraught with difficulties. The line between blame and accountability was not a distinct one and some anecdotal evidence suggest that even in the presence of the espoused theory, there still are acts that indicate that the industry is not really at a position where one could deem it to be characterized by a no-blame culture. Whether such a position is attainable or indeed required is debatable. The nature of professional accountability suggests that irrespective of what name is given to 'blame', or where in the organization (or wider system) it ends up being laid, there is in reality always a 'pointing of the finger' – the 'end of causation' within the bounded rationality and epistemic limits of humans – where things are perceived to have gone wrong. It helps in placing the accident in a particular context. It is what a company does – verifiably and consistently after this contextualization, this 'end of causation' (for want of a better term) – that would limit or promote the negative consequences 'blaming' has on learning. In other words, in any analysis of accidents, conclusions would be reached that unavoidably 'point the finger' at components of systems with or without attendant human intervention. It is the organizational actions that follow after the pointing (based on how organizations view the accident causation process), that determine how crew would subsequently contribute to the learning process. Where the actions of organizations (for perceived culpability) are termination of contract, suspension or even criminal sentences irrespective of context, ship officers will arguably never believe in the supposed 'blame-free culture' no matter how many manuals espouse the theory. Where blame is attributed and the consequences are retraining, system reconfiguration, and so on, ship officers are more likely to contribute to the learning process, irrespective of the term that is used to describe the 'finger-pointing' process.

While there is a growing recognition of the dangers of blame, the research indicates that this is not a wholly accepted view taken by even very safety conscious and relatively high reliability companies in the industry. Surprisingly and ironically and perhaps indicative of how entrenched a tendency to penalize is in the industry, one ship officer (at the sharp end) had this to say:

> As a team must be open to all regarding safety. Safety is best policy. Safety first before anything else ... About maritime pollution state that human error is about 70 per cent make it. My opinion is strongly implementation of penalty to those committed crimes. (48-year-old Filipino Chief Officer – Male)

Worker Engagement

Quantitative research findings:

- Mean worker engagement score was 3.713 (on a scale range of 1 to 5).

- Worker engagement is the highest predictor of organizational learning as compared to team psychological safety and leader inclusiveness even when controlling for rank, age and time spent in company.
- Management level ship officers had a statistically significant higher worker engagement than operation level ship officers.
- 'Income satisfaction' has a statistically significant main effect (but small effect size) on worker engagement. 'Mode of employment' and 'nature of employment contract' did not show statistical significant effects on worker engagement at the p. < .001 level.

The mean of 3.713 is moderately high, but still well short of a potential maximum. That, and given that worker engagement is a significant predictor to productive organizational learning, suggests that it is worth the while of shipping companies to focus more on worker engagement. Perhaps in recognition of the importance of worker engagement, the purposively sampled companies in the research had a number of ways for trying to increase worker engagement. In the context of this study the most important consideration was social relations on board and between ship and shore. The research shows that how credence is given to the views and contributions of the crew (ship officers in this case) is another important contributor to worker (crew) engagement. It is noteworthy that a substantial percentage (62.4 per cent) felt that their views were only sometimes considered or worse, that is, rarely considered, no influence at all or not sure. One officer put it very bluntly:

> [Companies learning from their seafarers]: That's a fairy tale! (30-year-old 2nd Officer – Female)

Given the amount of variance of worker engagement that predicts organizational learning, this must be an issue that the shipping industry takes seriously. Where views are considered, the officers indicated that companies consider the most important factor for credence-giving to be *rank*, followed in order, by *time spent in the company, external experience from other companies, informal relations, age, nationality* and finally *gender*.

With regard to the social relations on board the most significant approach used by the companies is the restriction of the nationalities they employ to only a few together with high retention rates. As has been mentioned, this has the effect of organizational acculturation and the blurring of national diversities. High retention strategies also make it possible for companies to consider social welfare packages that benefit their employees. These are not widespread in seafaring especially with regard to developing nations as a source of maritime labour. Olive Shipping, for example has a quite impressive but unique social welfare 'package' for their crew. Their organizational structure incorporates a position for a social welfare officer in the head office and a comparable position in the Philippines.

Andrea [Olive Shipping Social Welfare Officer – Europe]

So Olive Shipping is very unique. Yes it is. I think longer up in the system they may be certainly thinking about the money side of things, long-term, but also they have … Olive Shipping is very special in having prioritised the welfare side of things for their crew. Them having a welfare officer is something special in shipping.

[The most motivating thing for the crew] would not be wages because I've been told that wages are higher in other companies and I've experienced that people go, leave us, go to other companies to have higher wages but then come back again. My gut feeling is the whole environment. People stay here a long time. They get to know the people on board very well. They have sort of like a family there. And the welfare side of things – which I get to see a lot of – they are very well kept. They have good pension scheme, excellent pension scheme. They have excellent travel insurance. They have travel insurance for both them and their family. They have free time travel insurance – not just when they are going backwards and forward to the boat. They have a system, like, when they are sick that we follow them up; we can pay for this – for the operation. Sort of like those extra little things that mean quite a lot to people like I've been seeing. 'They see me as a person not just as someone earning money for them'. Yes. So, my gut feeling would be that. And they feel, some of them feel also – especially maybe the older generation – feel a loyalty to the company. The younger generation, I don't think that loyalty is much there. No. I think it is more that they feel that they're are welcome, they get along well with their crew mates, the wage isn't too bad, they've got a lot of social systems that they, schemes and everything that they find out about when they're sick and benefits and it's like 'oh wow maybe I should be here instead of changing companies'. So I think it is a combination of many things, but wage is not one of them.

[I have a Filipino counterpart]. She's sort of like my counterpart because she has this family matters especially and with insurances. So that's rather a new position really in the Philippines, because she has had more and more work, because the Olive Shipping Company in the Philippines has now been taken out of another company so like they are on their own now so they've made her position sort of like the one that I have. So we converse on different matters, like if I get sent a doctor's report and everything on board then I send them over to her for the Filipino's insurance cases and everything, I send all that to her. So she has, so we, when I was down in there sort of 14 days ago, we had a day where we discussed sort of like 'how do I do this and how do I do that and how are we going to work' like the different things that I can see that she would have the benefit of knowing about then I would send it to her. She's got the same sort of areas that I have but a little less because they don't have a health system in the Philippines like here, they don't have follow up of those who are sick in the same way. And the Filipinos they're engaged for their contract for sailing and then they are out again and then they come back again, whereas we have them all the time … I understand in Philippines in the Olive Shipping company there they have certain systems as well to try and hold the people that they have working for them in contract period like they have this welfare officer which they haven't had before. I know that some of the officers, I'm not sure if they are prioritised to come into training at the schools down there for, into the maritime schools with Olive Shipping. There was talk about they were going to maybe have a different sort of pension type of scheme for them as well. So they have different things going on there to try and keep the crew that they've got so they'll keep coming back to Olive Shipping. So Edith – the lady who is doing this – she'll probably be able to mention more about it because those special sorts of things we don't have too much in common.

According to Edith, professional development is also key to crew motivation and engagement.

Edith

They're motivated because in Olive Shipping the career growth is very, very fast. Because of, as you know, the scarcity of chemical tanker crew, right now it is a very, it's an alarming problem. I don't know in the near future, but right now there really, the percent, the scarcity is really very high. So here we are really pushing them, you know to go through ranks I mean to be promoted. We support their review and they know that in Olive Shipping if you perform and if you have a good evaluation, a consistent good evaluation you will be rewarded, that this good evaluation is always being looked at.

The detailed involvement of company in the welfare of the crew is not widespread among global shipping companies.

Other considerations to increase worker engagement were more geared toward creating an environment within which the crew are comfortable, aware of the companies' acknowledgement of their work and role and interested in their professional development. Salary is important and at least one respondent felt it was the most important motivator for crew.

51-year-old Master – Male

The most important food provision and salary rate.

However, in consonance with the more generalizable quantitative findings (ship officer views), it is not considered to be the most important factor by the companies (once it is perceived as fair).

Paul

My gut feeling is that there is no single motivator that will keep anyone [engaged]. OK, you need to have the salary otherwise it's not going to cut it anyway. But there is a difference between a competitive salary and a top salary. Olive Shipping is not a top salary company. So, I mean, for our Norwegian, Swedish, European officers in general, the salary is not what draws them to a job. We know that [laughs]. No we try to give them as much as we can but that's not what's keeping them there. It's the, probably I think the honesty and the openness of the company is a large contributor. Because I think if you didn't have that, I think there will be a less incentive to stay. The benefits on board I don't think they're that special. I mean we provide … they don't have internet on board for instance which is proving more and more a problem for the European officers.

Alexander

I mean if I don't know the company I would say this is because they are getting well paid – very good living conditions on board; which are very, very good – extremely good I would say. I don't believe that this is the main factor [for their engagement]. I believe that the main factor for them is the security they feel with this company. No matter what happens the company takes care of the people. It's the respect that everyone gets … The respect they get from the company, the security the feel within the company. They are treated, I would say from the onboard management as well as office management, with respect and fairness. After all we are all humans. We can make mistakes, but I believe we are fair and we are open. I mean anybody has a feeling, report something or claim something or complain for something, at least he will be heard not only by his neighbour; by everyone.

Human resource retention was a significant effort used by these higher end companies to increase worker engagement, incorporated as it is with job and social security and welfare dynamics. Other factors are a sense of fairness, transparency and opportunities for professional development. All of these point to the overall notion of being acknowledged and treated with respect. When the crew feel this to be the case, it appears they felt a sense of belonging and were more likely to communicate their own perceptions of risk and welfare as their contribution to the optimum performance of the company. According to Krause:

Getting safety right requires everyone's willing participation – and more than their participation, their active and wholehearted engagement. Whether such engagement comes easily or is almost impossible to achieve depends on the atmosphere leadership creates. The key to safety success is for leaders at all levels to cooperate to achieve the common goal of making the workplace safe.

We reject the dichotomy, sometimes assumed in discussions about safety, that the safety interests of labor and management are different and opposed. Safety improvement is a process built on explicit cooperation between workers and leaders of all kinds. Everyone has a role to play and the robustness of the safety improvement process is in direct proportion to how well everyone does his or her part. (Krause, 2005, p. 142)

Most of the companies try to encourage, create and nuture the environment where the use of some of the measures indicated would foster engagement in all that matters to the company including safety. However, the results of these methods remain intangible and frustratingly difficult to measure. In terms of practical engagement, and uniquely among the companies researched, Chrestos LNG has started the use of a behaviour-based safety system as part of a tangible formal safety system in a bid to increase the role played by shipboard crew in the direct management of risk. Behaviour-Based Safety (BBS) is part of the value-based safety system proposed by some researchers in the safety field (Krause, 2005; McSween, 2003) to increase direct crew participation and engagement in the risk construal and safety process. The key components are:

- 'A behavioural observation and feedback process
- Formal review of observation data
- Improvement goals
- Recognition for improvement and goal attainment.' (McSween, 2003, p. 17)

In brief. McSween summarizes the value of the process as follows:

Studies have documented the long-term effectiveness of an intervention package that includes an observation process, behavioral feedback, improvement goals, and reinforcements for improvement. The empirical data on the components of this approach suggest that an observation process, on-the-job feedback, and improvement goals are each important to maximising safety performance ... The data now clearly suggests the importance of involving all employees in conducting safety observation. (McSween, 2003, p. 18)

The system works by having crew look out for any behaviours they think compromise safety. This is then noted and fed into a formal review system. It is deceptively simple. McSween (2003) notes that 'many people underestimate the difficulty involved in creating a behavioral safety system', perhaps explaining why according to Alexander, many companies give it up. Chrestos, however, expects that the system will help achieve and maintain safety goals.

Alexander

> But we believe that the main factor that will change the situation [of improper reporting and no fundamental risk awareness] is behaviour. People must change their mentality … We have the BBS process. BBS. Behaviour-Based Safety. We actually introduced it, the system, last April. We started. OK this is only the very early stages. I am not sure that people realise how this thing works. I know that other companies that they have done in the past, they give it up. They are not actually taking advantage of this one. We'll give it a try. We believe that if we are consistent, in the long run we are going to expect results in a year or two years. five years and after, following this one, people's mentality will change. Because by recognising, let's say the mistakes that you do when carrying out a job, when I do the same job, it's a mechanism that works automatically. 'Oh, I said Manuel you were not wearing glasses when painting. What am I doing now? I'm not doing that as well'. And it works. It's proven that it works. You go and get your glasses. You don't do it like that because you feel irresponsible. If I made the observation on you and then I don't do it, what kind of example am I? And it works. But it takes time. It takes time.

The research confirms the hypothesis that worker engagement is a predictor of improved organizational learning and hence contributes meaningfully to the enhancement of quality and safety goals and aid in their achievement. As a result of these findings, Cameron et al.'s (2006) conceptual framework of worker engagement in Figure 7.5, can be said to be relevant to the maritime industry.

The shipping companies try to address the issue of direct worker engagement in two broad ways. The **first approach** is to try to engage the crew (ship officers in this case) in safety issues in a way that is legally mandated. The 2006 International Labour Organization (ILO) Maritime Labour Convention (MLC)[3] – Standard A4.3 paragraph 2d – requires that parties 'specify the authority of the ship's seafarers appointed or elected as safety representatives to participate in meetings of the ship's safety committee. Such a committee shall be established on board a ship on which there are five or more seafarers'. While the MLC is not yet in force, a lot of national health and safety laws under which ships operate, for example the United Kingdom Health and Safety (Consultation with Employees) Regulations 1996 (as amended), may also require this kind of safety representation for employees, in this case ship crew. Indeed this appears to be the most common way of crew engagement. The use of safety committees – along with the presence of crew representatives – was common to all the companies: Protection and Environmental Committee (PEC), for example in Olive Shipping. The ISM requirements for a Designated Person Ashore – who is to be a direct link between the crew and the highest level of the company (ISM Code paragraph four) attempts to create an engagement forum for the crew as well.

---

3   Not yet in force.

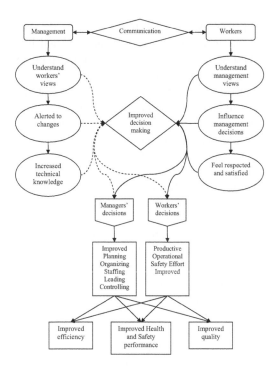

Figure 7.5   Conceptual model of worker engagement
Source: Cameron, I., Hare, B., Duff, R. and Maloney, B. (2006, p. 12)

In a **second approach**, the companies then had different company-specific and innovative ways to try and engage crew directly. The underlying principles were similar but the manifestations and emphases were different.

- As has been discussed in the section on Knowledge Acquisition, ship visits by senior managers (not only technical superintendents) were common to at least four of the companies. Significant emphasis is placed on this in BetaGas Carriers, Alpha Tankers and Chrestos LNG as required by TMSA to which they all subscribe. Informal conversations and ad hoc meetings are relatively high-value sources of crew risk perceptions for these companies. The value is enhanced by the increase in worker engagement that is reported as indicated in seafarer feedback.
- None of the companies researched use 'attitude surveys' or any other form of questionnaire solicitation for information on a regular basis. This is unfortunate. The results of the quantitative survey in this research effort indicates that ship officers do have a lot to say and it may be appropriate for them to have a regular forum where this is possible via questionnaires. To avoid monotony and questionnaire-fatigue a number of things are important

in this context. The first is that the frequency should not be such as to make such an exercise meaningless. Another issue would be the effect of the survey. It is best when the results are made known publicly and action seen to be taken as a result of the conclusions of the survey.

- Despite the absence of questionnaire-solicitation of safety information, the companies did solicit such information in most cases via anonymous suggestion boxes. The rationale is that the ship officers (crew) could send in 'sensitive' information without feeling threatened. While this may be helpful, in situations where it is extensively used by the crew, it does point to an absence of openness, fairness and trust and most importantly a lack of team psychological safety. In a number of cases this was the situation as evidenced by the picture anecdote alluded to by Jason (see earlier in this chapter under Team Dynamics).

- Safety circles (purpose-specific groups that bring together volunteers to solve particular problems as discussed in Chapter 3 in section about Worker Engagement in Learning) are not a very common practice in the companies. Perhaps due to the presence of safety committees and the fact that shipboard teams are already organized in teams that may be seen as project teams, the voluntary coming together of different crew members is not that common. Where this kind of engagement may be extremely helpful would be safety-circles between ship and shore-management. This, in the opinion of this writer would have a high worker engagement effect.

- Pre-task safety crew engagement is also a common feature. Indeed it is often required by the safety management systems. There is so much of this – via checklists – that it has led to certain problems, including a 'checklist mentality' (where it matters more to have the list checked and filed for audits, than it does to truly discuss and think through the subject matter of the list) and an undue burden from paper work. Where the task involved is not one of the many predetermined ones in the operation of ship (and covered in a safety manual and by checklists, for example), the companies create different ways of addressing the risks pre-task. In Olive Shipping this is called 'Task Risk Assessment'.

Paul

We also have task risk assessment which is basically any operation or any physical work – work from physical equipment I mean … [Indicates document] That is all the procedure for task risk assessment, when to carry it out, how to carry it out and basically how it is done is by way of risk matrices, so for let's say a typical task that should be risk assessed, crane operation, working pressure of vessel systems, work aloft at different levels, work outboard … For example you are doing something that is usually trivial and then there's bad weather. What do you do? Well you do a risk assessment before you proceed to do the work. You know the whole task risk assessment – the whole purpose of this is to raise the safety awareness ... On board, when they are doing an activity you know, you can't have procedures for everything which means that if there's anything that is unsure, anything that is not covered extensively enough by the procedures then you have the task risk assessment. So for tasks involving hazards that have not been sufficiently identified and controlled through already established procedures.

Unique to Chrestos LNG is a pre-task safety assessment they call the 'take two rule'. In the words of Alexander:

> And that is what we are telling them from here; briefing them and that goes for every body is the 'take two rule'. Before you carry out any job, no matter if you have done it two hours ago and then you break for coffee and you are back: take two minutes and access a few things. Simple things. You don't have to be a genius. Have I done the job before? Does anybody know that I am working here? What if I collapse; who will find me? What are the dangers in the job here – either from the job itself or surrounding or from the working conditions? And if all the answers are not clearly 'yes, I have all the answers', clearly – yourself, you don't have to ... don't do the job. If you have ... but be honest at least to yourself. If you don't have clear answers to your questions, the one that you raised, don't do the job. Go back and ask for clarification or logistics. If they do this, even let's say, once a day, but every day, every day, every day. This is how they will get used to recognising dangers, just like that. And the most you do it the most you are getting trained. We are asking the senior officers when they are, in the morning when they give the jobs to the bosun or junior officers, to have a little [talk] with them. 'OK this is your assignment for today; you have to go and inspect all the fire extinguishers all over the vessels'. Ask them. As a little check – but don't take two hours. 10 minutes. 'OK during your job, your duties, what are the dangers?' Ask them to identify the dangers. They will say a few dangers, either realistic or unrealistic. OK ask them 'What do you do about these dangers? How do you control them? How are you sure that something bad won't happen to you or to your assistant?' And this is the way that we try to achieve a higher level. These things cannot be documented because they are not measurable and we don't want people to look at it as a bureaucracy. They are already fed up with checklist and checklist.

A critical thing that appears to be missing is *an emphasis* on post-task discussions. This is also missing in the literature, but the merits – both for safety and for worker engagement, are self-evident.

- Reporting (of near-misses) is also a direct engagement scenario for crew engagement that the companies use. A few also use debriefing and exit interviews when the ship officer has finished a contract (three to nine months) and is proceeding on vacation. While this latter feedback is not task-based it does give valuable feedback on the whole safety experience of the officer and is recognized as a good engagement mechanism.

- A few of the companies use incentive schemes based on falling near-miss numbers or accident-free times (time between LTIs – lost time injuries – for example) to engage and motivate crew. Under certain circumstances (of transparency, honesty and blame-freeness) these have been shown to be helpful. However, as Budworth (1996, p. 24) notes, such measures lend themselves all too easily to suppression of information (and by extension, cynicism) especially where bonuses or such awards are involved.
- An important and more contemporary approach that is being employed by Chrestos LNG is the Behaviour-based Safety (BBS) as discussed earlier.

The discussions above show how the companies researched try to give opportunity (either as per legal requirements or as innovative safety enhancement approaches) for ship officers to be engaged in risk construal and safety learning on board ship. These approaches are shown in Figure 7.6. The underlined texts show the approaches that are most common and have the greatest emphasis in the majority of the companies. Texts in italics are approaches that were found to be non-existent or almost so in the context of this research and non-italicised/non-underlined text are approaches that are present to some degree but are not emphasized by the majority of companies.

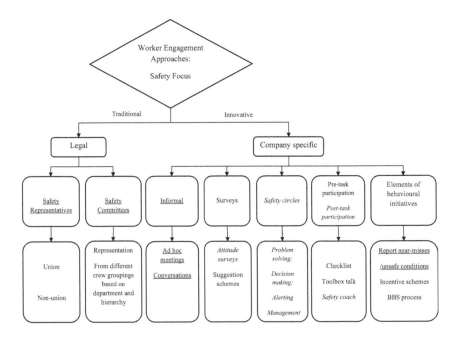

Figure 7.6   Worker engagement approaches: Safety

In these ways the companies try to create the opportunities for workers to be directly engaged in safety. Opportunity alone, however, is not enough to fully engage workers. Maloney and Cameron make this point very well:

> Involvement is a behaviour about which people make a conscious choice. A person can decide to be involved or choose not to be involved. As such, the critical questions are (1) what are the factors that influence a person's decision as to whether or not to become involved and (2) how do those factors influence that decision. This issue can be examined in the context of the following relationship:
>
> Involvement = f (Opportunity, Capability, Motivation)
>
> Involvement is a behaviour characterized by taking part in a process that includes activities such as evaluating a situation, analyzing alternatives, selecting a preferred alternative, providing feedback, and so on (Maloney and Cameron, 2003, p. 16).

To create the motivation for engagement is quite different from creating opportunity for engagement. The way the companies try to achieve a more holistic engagement of ship officers with the company and its safety values (as a subset of its overall goals and visions), goes beyond specific safety issues. These approaches have to do with the factors found in this research. They include crew retention, salary, social relations and amenities on board, team dynamics, contracts and crewing links with owners, professional development and social and job security and welfare. Figure 7.7 shows this more holistic background to worker (ship officer) engagement.

This comment by Cameron et al. is noteworthy:

> Management are required to collaborate with the workforce for the improvement of Occupational Health and Safety (OHS) on both legal and ethical grounds. However, these are not the only reasons for such activities. Interaction can lead to improvements in knowledge distribution and acquisition throughout any organization or project team. Feedback from workers can also be used to check management performance, increase productivity, efficiency and motivation levels as well as lower workforce turnover. (Cameron et al., 2006, p. ix)

Chapter Summary

See end of next chapter.

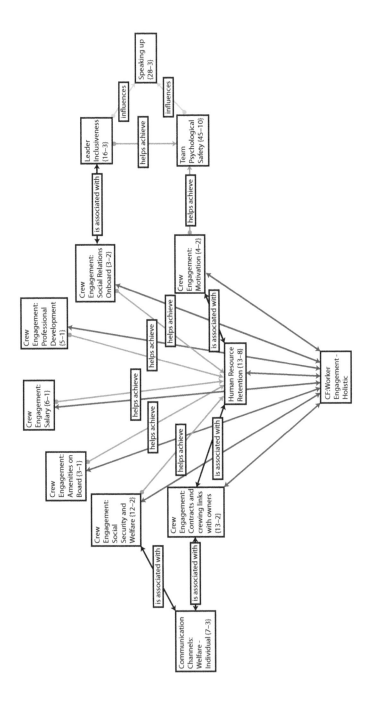

Figure 7.7   Worker engagement approaches: Holistic

Chapter 8
Discussion of Research Findings: Emergent Themes (II)

Past success does not guarantee future safety. Borrowing more and more from safety may go well for a while, but you never know when you are going to hit ...

Murphy's Law was wrong: everything that can go wrong usually goes right. And then we draw the wrong conclusions.

Dekker[1]

Purpose and Outline

This chapter continues the discussion of emergent themes as points of departure for the discussion of the findings began in Chapter 7

Impact of Accident Causation Philosophy on Learning

Some of the issues discussed have hinged on how organizations view the accident causation process and intuitively or explicitly model that process. Why are accident causation models important? A model helps organizations to think coherently about complex situations. It further helps generate common reference frameworks for communicating with regards to a particular context, and thus to allow for united (not necessarily uniform) and appropriate action to be taken across organizational boundaries. As Hollnagel (1998) indicates, the model held affects the choice of method used in accident investigation and prevention, the classification schemes derived for accident causation factors and the data sought, observations made and outputs created from analyses. The mental/theoretical models of risk, accident causation and learning inform processes of risk-associated data collection and analysis (Dekker and Hollnagel, 2004, p. 82; Hollnagel, 1998, p. 89; Mol, 2003, p. xix; Svedung and Rådbo, 2006, p. 31).

Models are metaphors that evidence the way we think. It follows that where an organization uses a flawed model, there will be undesirable consequences as far as safety is concerned.

There are many models on accident causation derived from different philosophies of how accidents are caused. Although they are often treated as if

---

1 From the book *Ten questions about human error: A new view of human factors and system safety*, published (2005) in London by Lawrence Erlbaum Associates.

they purport to capture all the dynamics of accident causation, they seldom do and are at best complementary to other models. Causation itself is an issue.

> Accident models and accident causation models involve two different areas of endeavor. The point is subtle, but in my view it is absolutely imperative to recognize the difference. Causation models purport to present cause and effects without identifying the phenomenon; no beginning and end of the phenomenon is indicated. Accident models on the other hand, deal *descriptively* with accidents as a process that has a beginning and an end, and the elements of that process. (Manuele, 2003, p. 171 citing a letter from Ludwig Benner)

According to the industrial statistician George Box (1979, p. 202), 'all models are wrong; some are useful'. One would qualify this to read 'all models – though not necessarily wrong – are limited (like metaphors) in their ability to tell the whole story or show the whole picture'. It is difficult to capture the complexity of risk and accident causation in a single model without losing the necessary parsimony of modelling: 'theories should not be more complex than necessary'.[2] Some models are more useful in depicting this complexity in simpler ways than others and to some extent their usefulness and applications are dependent on the theories that give birth to them. Over the years the aetiology and modelling of accidents have evolved from the accident proneness models through engineering models, interactive models, sequential models, epidemiological models, and systemic models with the contemporary emphasis on resilience engineering models (Dekker, 2005; Hale and Glendon, 1987; Heinrich, Petersen, Roos, Brown and Hazlett, 1980; Hollnagel, 2004; Leveson, 2004; Perrow, 1999; Qureshi, 2007, 2008; Rasmussen, 1997; Reason, 1997; Turner and Pidgeon, 1997). It appears that this modelling has been mainly academic and that the industry as a whole just applies intuitive models that do not necessarily derive from academic/theoretical thought, research or aspirations. The most dominant view is the sequential view and the emphasis on barrier analyses is high. While there appears to be an understanding of the multiplicity of factors that go into accidents and the influence of the system as a whole, the practice of accident investigation is still characterized by a structural decomposition approach rather than one that acknowledges the inherent dynamism and interconnectedness of open systems – one of functional abstraction (Rasmussen, 1997; Rasmussen and Svedung, 2000). A structural decomposition approach tries to break down accident causation to a sequential series of triggers and does not normally address the complexity of co-occurrence and/or tight closed-loop interaction. The whole is not just the sum of its parts. These important arguments in support of a more true system approach are not

---

2   Attributed to William Occam's principle of parsimony – the Occam's Razor – that theories should not multiply entities unnecessarily. *Pluralitas non est ponenda sine necessitas* (Plurality should not be posited without necessity). See also http://plato.stanford. edu/entries/simplicity retrieved 25 September 2008.

evident in the shipping industry. The arguments for a more functional abstraction outlook sees accident causation as more than a sequential failure or absence of barriers but of the interaction of complex factors, many of which may be desirable in themselves. Where the sequential approach is the dominant view, organizational efforts at accident investigation are focused on barriers that failed or were missing and subsequently prevention is geared towards more appropriate barriers. Safety (understood as more barriers) is not the same as reliability or resilience of a system. Organizations exist in a world characterized by entropy. This is not a statement of organizational recreancy, defined by Freudenburg (1993, p. 909) as 'the failure of institutional actors to carry out their responsibilities with the degree of vigor necessary to merit the societal trust they enjoy'. Entropy is characteristic of natural existence and functioning and is kept ongoing not only by organizational failure, complacency or lethargy, but also by changing external conditions irrespective of organizational performance.[3] This accounts for the 'incubation period' identified in some studies of accident causation (Turner and Pidgeon, 1997), a 'metastable phase' (Hale and Glendon, 1987, p. 14) where the dynamics of an accident in the making are gathering pace, irrespective of whether the organization notices this or not.

Considering that one of the benefits of having a model is to share and communicate common reference frameworks, the research found that the use of accident models is very limited in the shipping industry from an operational point of view. It is even questionable whether ship officers think of accidents in terms of any particular model or accident philosophy. It appears that the engagement with accident causation and prevention is very intuitive, superficial, reactive and on many occasions ad hoc.

Vasilis

Yes once an accident or incident will be occurred on board a vessel and then the captain immediately is sending us the accident, incident report which is a standard format, stating who is the person, what was the nature of the accident, when happened, if risk assessment took place, time, etc., etc., if he was sent to hospital, if he was medicated, etc., etc. so and all the relevant, let's say supporting documentation. This accident is analysed by the department. Also we invite other persons from other departments for example if it was let's say, if an accident took place in the engine room we call the superintendent engineer, or if an accident occurred during discharging we'll call somebody from the operations department. Ok. And we identify the primary cause and the root cause of the accidents. We are learning lessons and we relay these lessons – we disseminate these lessons under the shape of safety bulletins on board the vessels in order to see what happened and to avoid reoccurrence in future.

---

3   See next section in this chapter.

First of all we have attended – most of the superintendents have attended – safety, accident investigation courses. OK which are done by most of the classification societies like DNV, like ABS [American Bureau of Shipping]. So they have given various models OK and we are following such models but also we will have our experiences because we are doing this job for many years and we know how we proceed – how we are going to go through all these accidents in order to find the root cause, to find out lessons and disseminate to the crew of our fleet.

Yes, yes [we do have a standard model], which is not a strict model; OK we can, depends on the nature of the accident, we can deviate or if we meant more in detail, etc., etc., it depends.

Two of the companies – Alpha Tankers and Chrestos LNG – use the same accident causation schema – toolkits for root cause analysis that are 'borrowed' from a P&I Club and a classification society respectively .

Tim

Now an occurrence is accident, near-miss, technical incident, failure or something or other where nobody got hurt, basically piece of the ship or piece of equipment failed outside of the routine planned maintenance cycle, audit inspections, all of those are put together and reviewed and we use the same root cause analysis for all of them and apply that root cause analysis across everything ... As a management we break it down to three root causes, and we go: lack of standards, lack of compliance or lack of planning. Those are the levels that we break it down to at the bottom of our pyramid. Lack of compliance. Did we have rules in place that they did not follow? Lack of planning. Did they prepare properly for it in advance? And then finally ... lack of standards. Was there enough guidance for them?

Yes. The findings we then have, what went wrong? Why did it go wrong? What were the results of your investigation? And how do we fix it? How can we prevent it? Those are the three important things. That's what is important in any investigation. OK and then we have the root cause analysis which is basically ... this is the standard [refers to root cause accident causation chain in system manual] to which we have added an additional section, actually which we 'stole' from the UK P&I Club [laughter]. So you've got consequences, you know what were the potential consequences of the accident. You've got location. You've got what actually happened, the occurrence, the direct causes, unsafe conditions, you've got direct causes, unsafe acts so that's your personal factors, direct personal factors. You've got indirect causes of job factors.

When I am doing it day to day? No. Don't care [about the discussions in academia and the evolution of models of accident aetiology]! I honestly don't care! What I am, what I am considering when I'm doing an accident investigation – and I've done quite a few accident investigations in my time, some of them serious, some of them into near fatal incidents; in fact one of them was into a death but that was neither here nor there, and I don't care. What I am investigating is what went wrong, why did it go wrong, how can I prevent it. And that is all I care about when I am doing an investigation. The rest of it is irrelevant. Those are the three questions that I ask. The rest of it is fluff around it that everybody else in the industry likes to see, wants to see or would love to see written in the report but really what matters, really what it boils down to is what went wrong, why did it go wrong and how can we fix it; how can we prevent it from happening again. That's all that matters. You can categorise it, you can break it down and we do exactly the same here and we do it for statistical purposes, because we want to show oh look 80 per cent of our accidents in the world were caused by personal factors. No they weren't. No they weren't. 80 per cent of our accidents were not caused by personal factors. They were caused by systems that allowed personal factors to have an accident. Now if you're saying that 80 per cent of our accidents in the industry are human factors, 80 per cent of our accidents are deliberate. Because the only human factors accident that can really be caused by people is if someone does it deliberately. That's a human factor. That's the only human factor. Everything else is a systemic failure of some form or another in the shipping industry. We don't want to admit that but that's the reality of it. And I know that's not a popular view within the industry. These things help you get to those answers and that is the reality. And you step through. You've got to break down; you've got to have some form of structured investigation and the Swiss Cheese model of defences is as good a model as any; human causes, direct, indirect causes grid that we're using here, the DNV M-SCAT [Marine – Systematic Cause Analysis Technique] model – they are all the same thing. They're a formulated way of getting to certain areas to see what your defences were, to see where those areas failed and to give you an idea.and I think a lot of them stop in the wrong place. I think a lot of these models get down to this level – the root cause level – the basic level, the primary cause, whatever. Basic cause, primary cause, first cause whichever model you use and which ever model you choose – root cause whatever it is – and they stop there. And that's where it fails, because that doesn't solve your problem. All that does it tells you what your primary cause is, what the first point in the accident chain was. It doesn't tell you ways that you prevent it ... So this is the last one that we've put in [the lack of planning, etc.] which actually we took from – our Asian colleagues took from UK P&I [indicates the box in Figure 7.8 with the lack of compliance, standards and planning]. And UK P&I call it control actions. Here it is ... whatever you want to call it. It doesn't really matter. This is where you look to fix problems [taps the table in rhythm with words for emphasis] and that's the definition I've put in. What area should we look to prevent reoccurrence and these are ... I mean these are straight off the UK P&I model but they're actually a good set of things to look at and if I had the option starting from scratch this is what I would do. I wouldn't go for root causes, I wouldn't go for human factors grids, I wouldn't go for direct and indirect causes. If I had the option of starting from absolute scratch, I would have a very simple investigation process for any accident. I would require four things: what went wrong, why did it go wrong, how do we fix it, how do we prevent it. And I would give those 10 items as your guidance on where to look to fix it. That's important because that really ... We make things more complex than they need to be. DNV's M-SCAT is a classic example of an investigation process which is so complex. It's a process unto itself; it's a thing unto itself. But actually people don't understand it. 90 per cent of the people who use M-SCAT do not understand it. They just follow the steps because they feed you to the next step and the next step and the next step until you get to the bottom and it's now what it calls basic cause. Most of the people that use it don't really understand what they're doing with it. But I mean that's my own personal view.

There seems to be some confusion in the industry between methods and models, causation and processes. A model is a picture of accident causation dynamics: a philosophy of accident aetiology often expressed in metaphoric terms. A method is a formulated way of answering the questions that arise out of this philosophy. Though the methods (in the form of various tool-kits) vary, the dominant accident causation philosophy appears to be one that intuitively subscribes to the sequential view of accident causation. In the maritime industry there does not appear to be the debate that characterizes the issue in academia and to some extent in other industries.

The research participants' views suggest that the industry (shipping companies and their accident prevention and investigation processes) is significantly lagging behind both academia and other industries. Seafarers' contribution to the 'learning from accident' processes of modelling, categorizing and investigating is mainly limited to a narration of what happened. Even Alexander, who had noted that experienced officers were more likely to recognize more underlying issues, further noted that the categorization of incidents in the investigation and learning process was done in the office.

Alexander

For let's say, near-misses including these things, we [in the shore office] categorise them. We do not allow them [on the ship] to categorise them on board because sometimes they don't recognise easily where things fall. They see the obvious, but they do not see why. Behind the obvious what is the reason. So we categorise them, we circulate them to the whole fleet on quarterly basis, apart from being available in the database ...

Chrestos LNG uses the same model used by Alpha Tankers.

Alexander

I will show you [how the company views accident causation and the model we use for investigation]. Actually we have borrowed it from Lloyd's. I will show you. Have you seen it? I'll give you ... I found, I have seen some other models but I believe this one is very simple, understandable. Very, very important to be understandable by the people and then leading actually to three root causes which is lack of compliance, lack of standards, lack of planning. It cannot be anything else [that] can fall outside of these conditions; lack of compliance – which is the majority of the case, lack of standards and that's procedures, and lack of planning. And I have seen it on all the investigations that we carry out that it is [spot on] ... [The seafarers do not have the same mental model of accident causation]. They can see only the obvious. Most of them; they see only the obvious. They don't see the underlying factor ... I found it very, very good. I believe that people that they spend some time with it, they seem realistic. Makes sense.

The 'model' that is referred to by Tim and Alexander is shown in Figure 8.1.

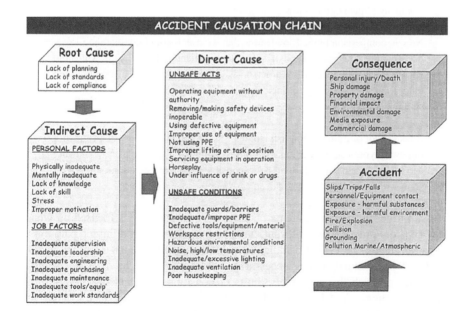

Figure 8.1   Accident model used by Chrestos LNG and Alpha Tankers for prevention and investigation

At one stage in an interview, it was pointed out to one respondent that there were other models that addressed the issues at a more appropriate systems-thinking level, while criticizing the linear thinking that dominates the model in use. His reply:

> Next week I will be in the United States from next Tuesday till next, next Friday. First few days will be on a workshop – safety – but the next week is going to be on, let's say software – it's called Reason [reference made to Wagenaar and Reason's Tripod[4] theory based on the epidemiological model Reason proposes (Reason, 1997)] and I am going there. I want to see but secondly I am going there to, and I will have this copy of my flowchart and say 'OK let's have a case, have a team. Let's run a case by Reason and the other team let's run by this one and let's see the outcome.'

After this respondent got back from the conference in the US, the following were his comments – sent in an email:

> Well the software itself is good and the process that uses to analyze an incident is actually of much value. The logic behind it (and to this you don't need actually the software) is of high quality. When it comes to Root cause, it gives you more than three even 10 root causes with a percentage per root cause contributing to the incident. Also the root cause is always driven to a point that states that the Company/Organization is always at fault.
>
> Now in comparison to the Lloyds flow that we use, I would say that the Lloyds one is more 'marinated' and it helps an experienced investigator while the REASON should be used by a person who preferably is not an expert in the field. REASON is mainly for incidents of a big scale and not for the majority of shipping cases.
>
> It takes a lot of training, English language skills and continuous use of the tool otherwise you lose orientation. I would recommend it in case that a Company can assign a team using this and be responsible to run investigations. Otherwise sooner or later it will become a dead case.

This – further to Tim's comments – shows how the industry is focused on what they consider practical and relevant, but which many argue is a limited paradigm or world view, at best. That paradigm uses a structuralist language that is limited to describing context and is necessarily reliant on hindsight. It is a pseudo-systemic approach that invariably limits or constrains accident investigation to static metaphors (models). Current accident investigation, as a result, tends to exclude the dynamic processes of drift toward the margins of safety. As Dekker

---

4   See http://www.tripodsolutions.net retrieved 14 May 2009.

argues, while reminding people of context is important, 'it is no substitute for beginning to explain the dynamics; the subtle, incremental processes that lead to, and normalize, the behavior eventually observed' (Dekker, 2005, p. 7).

The research trend in accident aetiology has progressed from reactive accident causation analysis to a concern about how to specify the systemic precursors to accidents (Dekker, 2005; Leveson, 2004; Pidgeon and O'Leary, 2000, p. 17; Rasmussen, 1997; Rasmussen and Svedung, 2000; Rosness et al., 2004). The shipping industry, together with their safety-training institutions, does not seem to be engaged in this enhanced thinking of accident causation. To be able to so recover and be resilient, an organization and its systems must have double-loop inquiry and learning processes that focus on system precursors inherent in normal work and not only in deviations or failure in a reactive mode. The difference in this context between a precursor and a near-miss and the advantages to be gained from precursor analyses has been already noted and discussed in this work. The industry – even the higher end operators as represented by the sample in this research – is still 'struggling' with trend analysis, with contemporary industry analysis being more a counting of accidents rather than an examination of fundamental patterns and co-occurring processes. This is not to say that this latter form of analyses should be subscribed to exclusively. There is a place for single-loop learning and a place for double-loop learning. There is a place for incident reporting, collection of near-miss data and a place for precursor analysis that focuses on inherent underlying patterns and the processes that insidiously lead organizations into adverse situations. In the view of this author, the industry should not see accident progression models as either-or scenarios but in the main, as metaphoric and complementary 'explanations'[5] of the challenging complexity that is normal work. Narratives of the anatomy of an accident may be necessarily linear/sequential/a chain, but such narratives should not be confused with the essential aetiology of accidents which is random/chaotic and dynamic.

Resource Availability and Utilization, Organizational Entropy and Drift

Research indicates that organizational functioning is characterized by drift toward the safety margin (Katz and Kahn, 1978, pp. 22–3; Rasmussen, 1997). This migration to the border of safety is introduced by what Mol calls 'entropic' risk – 'the degradation of a company's system factors (of) processes, technology, the physical environment and human resources' (Mol, 2003, p. xxi) over time. Shipping organizations, like all other organizations, are characterized by 'creeping entropy' or 'migration towards the boundary' of acceptable performance (Rasmussen,

---

5   Some of the philosophies that inform causation models are fundamentally different. Nevertheless, the metaphors or pictures used to describe the progression of accidents are mainly complementary.

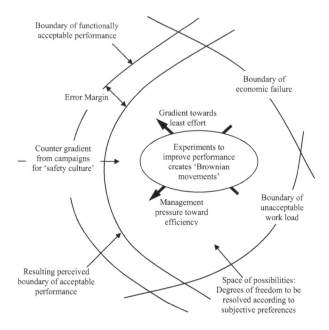

Figure 8.2   The migration of organizational behaviour toward safety boundaries

Source: Rasmussen, J. (1997)

1997). Figure 8.2 shows Rasmussen's conceptualization of the phenomenon of organizational drift.

Additionally, the drift can be due to a number of factors including complacency and resource constraints.

There is evidence of complacency that creeps in with many years of accident-free operation. This *Icarus Paradox* (Miller, 1990), discussed in Chapter 3, largely exists because the companies view safety as the antonym of risk and equate safety to the absence of accidents. Arguably, this is a position with questionable merit. 'Safety is more than the antonym of risk' (Möller et al., 2006). To illustrate, consider the difference between health and sickness. To say that one is healthy would imply much more than not being sick. There are in turn different degrees of health best conceptualized as the degree of human resilience against disease-causing influences. Similarly safety is the degree of organizational or industrial resilience (how far an organization is from the border) to factors that could cause undesired loss. The possibility that many of the factors that promote drift may be unknown is what makes continuous double-loop inquiry necessary. When safety is not seen this way, but is considered to be present only because accidents are non-existent and near-miss reports are (for whatever reason) falling, the company sets itself up for an Icarus experience. This certainly was the case with a number of the companies researched.

Tim

In terms of learning I think they, before I joined the company, 2006 was a disastrous year ... I think they were shaken to their very core. I think that before the incidents they were very complacent. I think they felt they knew what they were doing and they were very good at it and they didn't change, didn't want to find anything to improve.

Alexander

As I said, thanks God, we don't have many, many cases. This is good, on one hand, that we don't have serious or many cases. On the other hand – and this worries me – when things are calm everybody is relaxed. And working on these vessels is really relaxing. There are – provided that you know what you do – these vessels are not demanding; physically demanding. It's computerised – totally computerised. So everybody is relaxed, relaxed, relaxed, relaxed and we don't want that. We try to shake them.

Chrestos LNG tries to 'shake them' by unannounced visits to the ships among other things. Again the focus is on the ship crew and does not address organizational complacency. What is needed is a new worldview of risk and safety and the engagement with the fundamental processes of accident aetiology and not only the outcomes.

The second factor that creates drift and migration towards the edge involves limitations in resource. It is often the case that precursor conditions – the drivers of drift and migration to the edge – are obvious. Nevertheless, they are subjected to fluctuations in 'signal-to-noise' ratios, that is, they compete for attention and resources with other conditions that may seem more urgent and pressing. Because resource is limited and perceptions of the relevance of the information differs, the social stations involved either amplify or attenuate the risk signal by the commitment or denial of resources for example, of time, attention, money and training. The industry uses a number of triggers – 'those points at which circumstances fall below some level defined as satisfactory or acceptable' – which are primarily based on resource-allocation behaviour (Kiesler and Sproull, 1982, p. 548). The determination of acceptable risk is a decision problem , that is, it requires 'a choice among alternatives' a choice that 'is dependent on values, beliefs, and other factors' (Fischhoff, Lichtenstein, Slovic, Derby and Keeney,

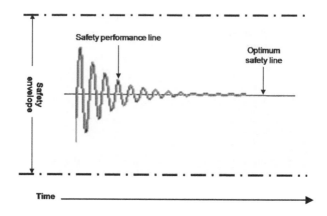

Figure 8.3   Ideal safety performance in an utopian world

1981, p. xii). The research finds that those 'other factors' are predominantly resource associated. The 'deviance' that becomes normalized (Vaughan, 1996) is not necessarily arrived at in quantum leaps, but are small minute deviations from safe practices – which are resource-dependent. Figure 8.3 shows the ideal progression for safety practice in a world of infinite resources and non-entropic

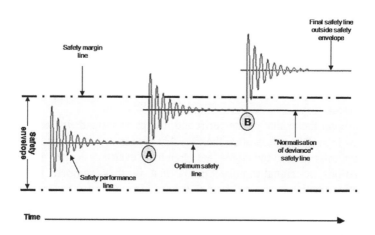

Figure 8.4   Micro-steps to the normalization of deviance in a non-utopian world

and static closed systems. Over time and with organizational learning the safety performance becomes synonymous with perfect and optimum safety.

In the real world, however, systems are open, entropy is a fact, resources are limited and conditions are dynamic. Figure 8.4 is an illustration of the incremental steps as an organization migrates away from the optimum safety line, first to normalization about a line within the safety envelope – but closer to the margin – and ends up outside the safety envelope.

The 'events' that start off new oscillations at points A and B, may be part of stochastic processes and are not necessarily big attention-grabbing ones. They may be micro-steps as individuals, teams and the organization adapt in behaviour, attitude or cognition to day-to-day pressures, internal and external changes and socio-technical, socio-political dynamics in the real world. The deviations do not appear as deviance and in the context of when they are implemented, are rational decisions that make sense and may even be applauded in the short-term. 'Companies today live in a very aggressive and competitive environment' which is focusing 'decision-makers on short-term financial and survival criteria rather than long-term criteria concerning welfare, safety and environmental impact' (Rasmussen, 1997, p. 186). It is only in hindsight that they appear flawed and that the sequential steps to disaster are clear.

30-year-old 2nd Officer – Female

Nowadays, everything has to be done fast, because time is money! There is so much pressure from above [Office, Client]. Very often there is no time to think about safety or pollution in a proper way.

36-year-old Chief Mate – Male

The company's cardinal objective is profit and all new innovations are judged in that light.

One particular company's use of a management company, which appeared not to be as conscientious in safety-related matters as the parent company was, had led to a number of accidents. However, it was acknowledged by the parent company that, though the particular subsidiary company involved in the accidents was no longer being used, the continued use of such external third parties in management of ships was an economic decision, significantly affected by the economic goals and contexts in which the company operated. The companies articulated in very clear terms the restrictions that economic considerations place on optimum safety investment. In one case the company had just been asked to pay taxes which were earlier waived in a government drive to increase the nation's flagged fleet. This new tax was to be paid retroactively at a time of global 'recession' and high bunker prices.

A respondent

Among other things the, you know about this, the Norwegian government with the new tax regime where they ... well ... the tax that we were not supposed to pay until we took out the money got reactively [sic] changed and so we suddenly have to pay, I don't know, 450 million or something and reactive [sic] tax to the government. The deal that they made, you know but that leads again to, you have to have more liquidity, you know, for the company at once. So suddenly you have a problem, you know. You have these major expenses to pay out and a dollar that's falling; you have, you know, the whole financial situation and the bunker prices that went up about 150 something. So this whole thing leads to, you know, a change in the fleet ... what you want to have internal-external.

The company felt that in such a context, minimum compliance with existing safety regulations would naturally be the outcome, showing that in a competitive environment, the perceived commercial benefits of operating at the margins of safety takes precedence over more quality-driven safety performance. This pressure to lower safety standards under pressure is not characteristic of only management in companies, but also operators at the 'sharp end' who sense such pressure and respond to it, despite the companies' overt statements purporting to discourage this.

The management of risk is replete with references to 'acceptable risk' and like phrases (Fischhoff et al., 1981; Manuele and Main, 2002). Indeed there is pervasive use of the concept in the IMO with the popular risk phraseology 'as low as reasonably practicable' (ALARP) (see for example, IMO 2002a) and the ISO/IEC Guide 51 affirms the existence of residual risk and tolerable risk (ISO/IEC1999). What factors, then go into the determination of the level at which risk is acceptable? Quite apart from the sociological factors of context, perspective, knowledge, experience, power and culture that have been discussed in this book, one other emergent theme that affects the 'acceptability' of risk is resource availability and utilization. The dominant role of profit-making in shipping is obviously pervasive and often perceived as being at odds with safety goals. While there appears to be a gradual recognition that this may not be the case, and that safety goals can and should be compatible and augmenting of productivity goals, the motivation for resource utilization is still very much based on productivity.

Alexander

If the control measures are too demanding, or too much to be implemented then we change the task or the job or whatever and we do something else avoiding the risk. However, all the risks are manageable. It depends on how much resources you are willing to allocate to have a risk down to an ALARP stage. And of course this depends on growing the company, resources from the company and time. When we talk about resources people think we are talking only about money, or even human resources. They are forgetting about time. The money may be there, the human resources may be there, but the time sometimes – in many, many cases – is not there. And this is something that people tend to forget.

Given the companies' acknowledgement of the imbalance of resource allocation to the investigation of near-misses as compared to major accidents, it may seem that the allocation of resources to precursor analysis will entail a 'quantum leap' that the industry is not ready for.

This brings to the fore the **balance between productivity and safety**. The two are perceived in the literature and in practice as being incompatible goals.

Alexander

It was not long ago that all companies were mainly focusing on performance. So as to match my [economic] performance, people used to take shortcuts. Now since a couple of years – a decade I would say – since ISM existence and implementation, asking the people to come to another mentality where safety is priority number one as it should be from day one, but it wasn't. It takes some time. I mean we have some way to go, not at Chrestos LNG purely but as an industry as a whole. And yet there is not a clear balance between to what extent we take the risk assessment vis-à-vis the productivity or the performance. It has to be well balanced. If you carry out on board the vessels, risk assessment for every little thing, then there is no space for any business on board the vessel.

Dekker, discussing the research convergence on some commonalities of the causes of drift, notes that 'at the heart of the trouble lies a conflictual model: organizations that involve safety-critical work are essentially trying to reconcile *irreconcilable goals* [emphasis added] (staying safe *and* staying in business)' (Dekker, 2005, p. 5). This mindset is pervasive – both in academia and industry. In as much as it is very real as a socially constructed artefact which influences all organizational decision-making regarding resource allocation and prioritization, it is worthy of attention. As long as academics and practitioners industry-wide view productivity and safety as incompatible goals, resource allocation will be seen as being limited to *either* safety *or* productivity. That is the way humans have made the world they live in!

Alexander

Industry wise is that even if I spend, as a company, billions training and developing people, the market won't recognise. If your vessel is cheaper by 500 dollars per day, they won't take my vessel, while I have the best crew. This is a reality and I can give you examples and historical examples of this. So this is the industry part.

Arguably this need not be the case. Productivity and safety can be seen as compatible goals or at worst reconcilable goals. Mol argues clearly for the possibility of such a worldview and describes the mechanisms by which practitioners can arrive at those ends (Mol, 2003). There is some indication that the shipping industry is gradually accepting that there is no real incompatibility between high safety goals and high-end efficient and profitable operations (Dalan, 2007). This is particularly true of the tanker sector where commercial pressures – which are more routinely a source of deviation from optimum safety standards – are increasingly aiding to combat the tendency to cut corners. The detailed standard-setting of TMSA – emanating from the industry body OCIMF – is evidence of this. The research clearly indicated that there is a perception that tankers are much better operated because of the recognizable risks and the impact of relatively stringent auditing and inspection schemes from clients/charterers especially the oil majors.

Paul

But I mean the oil majors have higher requirements than the ISM does. There's no regulation stating for us to have all the things that we have. A lot of this is due to the oil major's requirement.

Vasilis

[Safety on tankers is much higher than on the dry trades]. This is definite. This is definite. It comes out from statistics, from the various bodies and I think, my opinion, this is something which is discussed very openly and very widely. Tankers, because of the nature of their trading, because of the frequent inspections from the oil majors, because of the TMSA which was recently launched, they have improved the degree of quality ships, of quality crew and consequently quality inspections on board and have contributed to better jobs.

Evan

I am not sure that this type of [safety] culture is as strong in the dry sector yet. And that's because their customers haven't demanded it. The problem is that our customers are oil majors and if they don't see this, they won't do business with you. That's plain and simple. So it's been customer driven for sure.

It is interesting to note that Evan casts the role of the oil majors, in demanding higher safety levels for business, as a 'problem'.

Another benefit of regimes like TMSA that ties safety to commercial productivity is that it gives safety managers access to resources in a manner they would not have otherwise.

Evan

Again I think one of the things for us is that we, ourselves and the, you know in the tanker companies, we get audited a lot by the oil majors, you know and other people who are much, much tougher than ISM. I mean ISM is the absolute minimum, you know standard. And of course that makes us, that focuses us. So I mean, I tell people over and over again, I think TMSA is great because it gives us a clear road map of what our customers want at the end of the day. You may argue a little bit about best practice, about what is best practice, but the KPIs [Key Performance Indicators] are clear OK. And when I sit down with my commercial friends and they want to know why do we have to spend money on that; why do we have to do that and for me a very, very easy answer is 'it's in TMSA' [laughs]. And, you know, at the end of the day, you know, do you want to do business with the oil majors? And if you don't want to do business with the oil majors, I can tell you, you know, all you have to do is not do these things [laughs]. Because if you don't do them, believe me there's plenty of companies, you know that operate vessels that will do them. Your challenge is, I always say, as a commercial person, is to get enough money to implement these changes and some of them are very expensive.

As welcome as the role of regimes like TMSA is, there is also a recognizable trend towards these charterers treating the inspection and vetting regimes not as a fundamental paradigm shift of safety as a culture (and the viewing of safety as compatible with productivity), but as 'insurance' against liability in the event of an accident. For the carriers, the quest to achieve the 'noble' objectives of the approach are seldom motivated by a fundamental belief in the compatibility of productivity and safety goals (the desired paradigm shift) but are driven by legislation (hard and soft) and a compliance mentality. This view means that where standards can be lowered without 'repercussions' like failing audits, losing certificates, losing customers, the company slides into 'evasive mode'. When one respondent was asked whether the company directly correlates high safety standards to high productivity, the 'interesting answer' was: *Depends how cheap they can get away with. I'd say, personally* [laughs]. *But ... at least I believe so. Yes.*

The new 'safety culture' that is developing in the industry may not be premised on a new view of safety and accountability, but on legal and commercial pressure. This is less than ideal, particular in light of the so-called financial crisis of 2008, which suggests that with restriction in resources will come the increased potential for cutting safety corners.

Paul

Personally I don't really think [that if there wasn't such intensive external commercial pressure for safety], the company would still do these things? No. I think the right approach to actually getting the companies to do this is to make it a requirement and they are profit – I mean shipping companies are profit organizations and that profit will always lead to goals being to earn more money.

> At the same time Olive Shipping as a company tries to keep a very high or very good profile. We sell on the fact that we are a good company. We are safe, and we are open towards our customers. So in order to be that company, we also need something like this. So from a selling point yes, maybe we would still have it. Personally I don't think if no oil major or no international requirements were in place, I don't think it would be [used].

Paul's statements bring into focus the role of legislation in safety management. As has been mentioned earlier, legislation is ubiquitous and indispensable. However, where a safety culture is driven by legislation, there will be problems. As Schuck notes:

> Law appeals to self-interest mainly by creating behavioral incentives that typically take the form of penalties or, more rarely, subsidies. Such incentives generate for law only the grudging respect that fear or duty elicits; this in turn supports a perfunctory, minimal compliance. Laws that must be implemented by administrative officials make self-interest doubly problematic as a motivational force; the self-interest of either the enforcer or the regulated may stymie the law's implementation, as often occurs in criminal and regulatory law. (Schuck, 2000, p. 437)

Law has limits and is only one agent of social change. Other forms of social mediation are available, for example social norms (acquired via education – formally and informally) and markets (which Paul mentions in the latter part of the quote). The maritime industry's extreme reliance on legislation means that most learning and change are compliance-driven. The sustainability of such change is debatable. The view for 'more law', more prescription, has recently been seen in the call for the ISM code to be made more specific and prescriptive in detail. ISM is by nature a generic code – based (in the words of the code) on 'general principles and objectives' (IMO 2002b, p. 5) and expressed in broad terms. This 'goal-based' approach is deliberate and in the opinion of this researcher, rightly so. Further specificity and detail would unnecessarily increase the prescriptiveness of the instrument. Such prescriptiveness will only in turn increase the administrative burden and erode further the emphasis on professionalism that all in the shipping community should strive for (see arguments (for and against) made by Anderson, 2008). Shipping companies and the trade groups they belong to should create the detail as in the TMSA. As such, calls for a global blueprint or standard ISM-compliance model, though well meant, may be misplaced.

The companies in this research sample are high-end operators and they are also (at least in part) motivated by the informal norms and their reputation in the 'marketplace'. This is an example worth emulating across the industry.

Vasilis

> We maintain a very good reputation amongst our clients and we've got to maintain it. If one time we have one problem; the other time we have next problem or other problem then in these cases we start losing face.

Another motivating factor for bridging the perceived productivity-safety dichotomy is the media. Companies recognize that bad safety-related publicity is detrimental to productivity.

Paul

> I think because of some of the accidents that we have had; take for instance *Atlas Arrow* which we've mentioned several times. *Atlas Arrow* was not the best of publicity. You don't want that kind of accident. So in order not to have a chemical tanker or tanker explode or something as well as a vessel that comes crashing what not, you want to keep it floating and you want to keep it working. That in itself will be a good motivator for having right procedures, etc.

To align organizational safety goals to productivity goals, company assets – both liquid and fixed – should not be the only indicator of organizational performance. Focusing only on assets entrenches the mindset that productivity goals are incompatible with safety goals. When the society at large and shipping companies in particular place more value on other intangible assets – the environment, human life and dignity – it will become more obvious that productivity, safety and environmental protection both in the short and long-term are not incompatible goals. Share-holders will see themselves then more as stakeholders and look for efficiency and results not only on the financial bottom line.

Related to the issue of productivity versus safety is that of the balance of ship operation interest between the owner, manager and charterer. In some of the trades the charterers take an above average interest in the operations of the ship.

Evan

> Yes [in LNG and other tankers there is significant input from customers]. That's true. I mean I think you do see the, definitely on the LNG side and the oil side, you know, the risks to the charterers are very high and therefore they try to mitigate those risks by understanding how the ship managers go about hiring their people, training their people, this and that. Some of them get deeply involved. Others, you know, not so much – don't get so deeply involved.

Sometimes this nexus of charterer/operator/owner has to be treated with caution.

Evan

Let me give you an example of a management of change that we're working on right now. We've been requested to upgrade the software on our gauging system for our cargo tanks by one of the charterers. Under the time charter they have the right to ask us to modify the vessel to meet their business needs. But we have to be careful with that kind of a change to software and things like that, to be very careful with that. And what worried us a bit was that the charterer had done all the discussions and, you know everything by themselves and then out of the blue, we didn't even know it was going on, they said 'oh by the way, you know, we've arranged to have this software created' and I said 'whoao, wait a second here, you know, you know that's not the way we do it. We're the shipowners and managers and we're happy to hear your proposal'. So what we did was we got in touch with the manufacturer. They came in last week, they gave us a full briefing on what the changes were, what the risks were, what the mitigation to the risk were, whatever. And now we're in a position and we have a form that goes with this thing that has certain levels of authority to sign off and make the change.

Safety considerations, risk construal and learning should have not only the seafarer in the loop, but also other third parties who are stakeholders in the safe operation of the ship. Where no such communication is evident, the most critical safety issues appear to lie in transferring of owner safety values to the entities which daily manage the ships (where these are different companies).

The point is reinforced by a respondent:

41-year-old Master – now pilot – Male

There are discrepancies between the way that a daughter (managing) company acts and the mother (owning) company [in reference to two corporate entities which are not congenitally linked] – different values which may make life more complicated.

For example, at the time of the *Atlas Arrow* incident (in which there was significant loss of life and the total loss of the ship), Olive Shipping as owners had contracted a managing group (referred to as Bedros in this text) operating out of Greece as managers of that particular ship and eight others. One of the interviewees (Edith) worked at that time with a Philippines-based management group (referred to here as Matrix) serving as manning agents for Bedros. At the time of this research, Edith is now working for Olive Shipping.

Edith

Ok. I think during the *Atlas Arrow* accident Olive Shipping didn't have – like me – I was with Matrix before and my principal is, when I was called in, Bedros; [there were] nine vessels belongs to Olive Shipping and Bedros is managing it. So my principal is Bedros. I am adhering to what Bedros tells me and I don't.

Olive Shipping doesn't say that 'OK this is the form that we want to be used, this is how the safety training should be done'. So it is Bedros which is doing like that so I am talking to Bedros and with that one I have the impression that Olive Shipping in that particular case doesn't have say when it comes to safety because of course this is being managed by Bedros, so Bedros is implementing their own safety measures, implementing their own Safety Management System on board so it's not Olive Shipping's that is being implemented. All the standards forms, all the, everything, hot work and hot work procedures, everything, all procedures is Bedros and not Olive Shipping procedures. So that time it is Bedros procedures that are being followed. So that's why maybe they end up not together anymore.

Paul

What the *Atlas Arrow* accident proved was that the follow-up onboard and the control the owning company had over the vessel was faulty: little or bad. So in a sense it proved that you need more control over your time chartered fleet or vessels on external management. So I mean, now you have, we have personnel dealing only with the external fleet, etc., the vessels, etc. But most of our vessels are on our management so ... [It is not Olive Shipping's goal to manage 100 per cent] No. I mean, it will depend daily on what is in the market. So it is hard to tell. You know, for the year that I have been here, we have sold and bought vessels but yet we are still around 50, have been for as long as I've been here. So, you know a certain percentage of the fleet yes, but it's not the percentage in itself that is the goal. It has to be the profitability, etc., for the vessels and the company.

While Olive Shipping terminated the managing contract with Bedros, none of the people interviewed were sure – in concrete terms – of a fundamental addressing of the safety control relationship between owner and manager.

Edith

Olive Shipping still have some other vessels managed by other companies but not Bedros any more ... To be honest I don't know if at the time after the incident of *Atlas Arrow*, they have gone through how they are going to interfere with the system of the managers of the vessels. I don't know if they have done something but I guess they have to and tell them what is, when you are managing our ships this is what memorandum of agreement should be. I don't know. Maybe, maybe but I think, I guess it should be.

Competence and Training

Most of the issues discussed hinge on the people involved having different long-standing and persistent perspectives. To their credit the organizations visited seem to place a high emphasis on training and the improvement of competence of their human resource. What one would postulate is that the kind of training – more rightly, education – that is needed is not pervasive enough. The companies focus on and have invested substantially in technical training. From a learning perspective the most noteworthy practice is how the lessons learnt are communicated through training. Some of the companies seek to complete the loop of learning by injecting 'lessons learnt' into the curricula of the training centres.

Edith

We discuss [risk concerns] in our management review meeting and coordination meeting between the department heads and then if there is something that has to be attended by the academy, by the training centre we send to them by means of telling them that this has been noted and you have to look into this and find out how we can put this, incorporate this one, this matter into a module or the course that they are having there. Maybe they can provide a new module or maybe they can just incorporate it in an existing training course. So I think it is OK.

Evan

We also then send those [incident reports] to our training centres and make sure that the training centres and the instructors on board are current with the current problems, you know that we are having on board so that when they have the people in for training, you know they're also dealing with relevant and updated problems that happen on ships.

For some this meant courses structured by them but delivered by external institutions. Others have gone so far as to invest substantially in what are essentially training academies competing successfully with traditional METI.

Tim

You know, we're spending – and quite rightly to be honest – we're spending millions of pounds, investing millions of pounds or millions of dollars in ships. We should be getting millions of dollars in return for what we do and we should be spending our money wisely and carefully. So investing in training is something that you have to take as an investment. You have to show, you have to accept that it's a long-term investment that you're doing with these people.

Edith

Olive Shipping has invested a lot of, lots of investment like as you see we have training, a training centre here, our own academy. We have cadetship programme where they have built their own campus in tie-up with local – with one good local school for maritime studies here. They have their own campus already. And then this training in Subic in SBMA [Subic Bay Metropolitan Authority – Philippines], they have put a lot of investment there … I think billions of investment they have put in there because it's a real, like this training centre is like a real ship. They have tanks, they have everything. All the equipment on board a ship is there. So this is quite a lot.

The 'feed-in' from 'lessons learned' is often very technical, being derived as they often are from the incident-report databases, which themselves are substantially made up of technical details because of the level of detail of accident causation philosophy and investigation models subscribed to by the companies. They exclude 'feed-in' that gives some attention to more theoretical and sociological aspects of risk governance, new paradigms and worldviews of accident aetiology and modelling and a new appreciation of organizational learning. This gap is acknowledged.

42-year-old Master – Male

Crew needs theoretical education. Work is based on practices very often. Crews are not educated to additionally cope with new regulations.

Alexander

Although they have the liberty and the right to speak and stand up and say 'there's a risk there that is not identified or we haven't taken measures to deal with it', crew don't do it. They don't do it. Not because they are not allowed, or because somebody will look and [question them], they don't want to take the hassle. I will equate it to laziness or somebody will call it lack of training. Identifying risks – apart from the obvious risk – is not easy. It's not easily identified. It takes training and well experienced people. Normal people who are involved in the work, unless the obvious risks that overhang them directly; they do not recognize situational risks. This is I would say a gap, not only within Chrestos, but within the shipping body.

To further address the softer issues of team dynamics and communication between ship and shore offices, companies showed a tendency to have more informal company seminars and conferences geared towards their own needs. For logistic and location reasons this is not mandatory and not all officers make it. However, all the companies which do this (perhaps only a few globally) are confident of the importance of this measure in increasing cooperation and establishing informal links that make formal links work more smoothly.

Andrea

We have had an officers' seminar, conference just recently in November and then they had invited both the senior officers and the junior officers to the same conference. Usually they have two different weeks. And that is actually a good thing because they get to mix in a social situation which isn't aboard the boat. They get to know each other in a totally different way, they have work things during the day but in the evenings it is relaxed; it's a totally different atmosphere – they've got good time and they don't have jobs they have to look out for. So social activities or courses where you mix both senior and junior officers, I think is a very good thing. They get to know each other in another way. Like it's not just Captain this and this but it is person this and this. So I think those things are actually quite good. So Olive Shipping has some of them but maybe the courses could be even more – or the conferences maybe they could have more sort of team-building conferences rather than conferences that had just a focus on the technical sort of side of things, but maybe they could actually have team-building things; exercises – because that is something that is not being done but how seriously it will be taken by the sailors, I don't know.

Apart from the issue raised by Andrea's call for more structured team-building 'syllabi' in these conferences, it is also noticeable that the conferences are mainly region based. There are good reasons for this including the scarcity of resources in terms of time and money.

People from different regions only attend when they are close to where the conference is taking place and it is mostly made up of addresses by senior management officers:

Edith

[All the officers (North West Europeans and Filipinos) meet together in one forum]. Yes. Although of course those on vacation are being invited to attend and of course programme is being prepared for them and then if there are Norwegians around – because some Norwegians are having training here in the academy, in Subic, in SBMA, they also being asked or invited to attend and to think something from their point of view ... I mean this is a normal regular process that we do, this management workshop ... What we also do is to ... that is one thing and then what we also do is to, whenever we have this management workshop we try to invite as many senior officers, like shore officers from Bergen and from Singapore, from the ship management group of Olive Shipping to attend and to hear what this management has to say during the workshop. And then of course if there are some Europeans having training here at the time we also invite them to attend.

Evan

The Magna Group holds a safety conference every year – an annual safety conference. In fact one's coming up next week, over in Eastern Europe and they hold one in the Philippines, they hold one in Mumbai in India and they hold one in Europe.

Jason

We have a seminar in Croatia every year … I think last year was in Dubrovnik. I don't know where they're having it this year.

Everybody (comes) … all the officers like, it's mainly Europeans. All the Europeans, they'll come there. Anybody that's on leave. Filipinos, no. I mean … they have … this might change. I mean we might have … they have their own one out in Manila … and we usually will send somebody there to give a lecture or something like that because we also do the ones here. The ones in Europe – we do, we go and have speakers there as well. So they meet everybody. They have personnel. Maria [shore staff] whose not here – she's travelling at the moment – Maria goes, John [shore staff] goes, I think Dima goes – will probably be going. We also have an American colleague – Bill – and Bill he's our sort of representative in America. He does a lot of ship visits for us, assists us with the coast guard … other stuff like this. He's got a lot of contacts in the coast guard. I mean when we've got, you know queries and anything with the coast guard we can go through Bill. So he's going to be going to the next one. I think it's in April actually. I'm not going. I might be going to the next one. That's when all the people from, in Europe, all the European officers at home on leave are invited. Some of them go. They use to get, probably about 100 people.

It is feared, however, that this kind of regional based training further creates disparities between members of team when they come from different regions of the world. In a few rare cases, there had been attempts to get attendance to the conferences from all regions. Andrea refers to one anecdote showing how some issues could still pose difficulties.

Andrea

But I think we have discussed with this team-building and teamwork conferences or meeting points. But we feel like at the conference now we have four Filipinos officers at the conference but all the conference was, the overheads were in English but everyone spoke Norwegian [laughs]. Yes. Yes. So you have sort of like those things that happen that you think, well here they are sitting and if you've got only overheads, fine they can follow it, but if you've got one overhead and you talk in Norwegian for an hour at a time what is the point of them being there? They have no, they work together [amongst themselves as

Filipinos] most of the time and they didn't interact with the others so very much [it actually reinforces the 'grouping' phenomenon and deepening distinctions]. Yes! So those sorts of things I think in this global company we have to maybe be a bit more aware of and maybe able to do things but then you have to have also the time and the prioritising of the money to do something like that.

The use of such conferences in the companies visited evidences the emergence of a philosophy of training that seeks to be determinate in the acculturation of the seafarer to the particular organizational culture of the shipping company. This has been perhaps intuitively always the case. What is different is that the process is being made more explicit (at least in these companies) in a bid to influence the seafarer cognitively and affectively irrespective of the existence or non-existence of cultural biases based on nationhood and to reduce the potential negative effects of rank differentiation. It is premised on the realization that 'culture' (specifically in this case, organizational culture) is dynamic and that many other influences can impact on it and successfully blur real or imagined distinctions of nationality and adverse rank distinctions. Partly because this kind of training addresses company-specific issues, traditional METI seem not to be well positioned to give this training. However, there is also a feeling that the continuing development of human resources require a lot more than METI offer – hence the substantial investments in company-owned and operated institutions and increased shouldering of the training 'burden' by companies. This trend has some disadvantages.

Alexander

The other part has to do with us in the company is that the way the things are taking, we are turning from a ship management company to a training centre. And people forget that when a seaman is on vacation, he needs to meet with family, with friends, take some fresh air, not to spend his three or four months on vacation on schools and coaching and speaking. It's a very, very hard situation.

The placing of more 'training' responsibilities on shipping companies would suggest a deficiency in the work of traditional METI or could be due to a lack of communication between industry and METI (academia). Generally speaking, traditional METI appear to be completely out of the loop when it comes to current and dynamic risk information distribution. None of the companies researched had *formal* 'feed-in' systems into traditional METI where they could send recommendations for curricula upgrades based on lessons learnt. The lines of communication were between the companies and the training centres they owned or in some cases bodies such as P&I Clubs and/or Classification Societies.

Alexander

We have a very structured training matrix, that all of our people have to deal with and a refreshment training matrix. And we deal with some respected training centres here in Piraeus, like ABS, Lloyds and some private ones, according to the subscribed ones, not only on the practical, I would say, trainings like simulation, but our focus is on the, let's say, higher degree, I mean, leadership, behaviour-style management, these things. Because we believe if we catch these areas, these hard to catch areas, then we can easily change the obvious gaps. But as I said this takes time. Takes times and we are on the way. On the way, but I wouldn't say that we have achieved this as yet. For the people that are within this company for many years, the bulk majority of the officers, things are better because they know what the company is demanding, they know how we want to conduct our everyday business. But as I said this Chrestos is new – LNG part. So we have employed a lot of new officers, trained them – technically trained them onboard, but yet, to move the challenges and the requirements and the philosophy of this company it takes time. They have to stay with us five service years to understand and realise what are the values of this company and what is our primary business.

P&I Clubs and Classification Societies are playing an increasingly dominant role in seafarer training. This may be because they have a direct stake and interest in how well a client (the shipping companies) performs. While the traditional METI, for example, offer Bridge Resource Management (BRM), it is a P&I Club which is at the fore of the introduction of Maritime Resource Management (MRM) (see for example Hernqvist, 2008, pp. 30–31 describing the MRM designed and run by The Swedish Club). Perhaps METI – globally speaking – need to establish better communication lines with industry and develop features that permit them to respond more dynamically to the learning needs of the industry at large. Tim's comments in this regard are insightful:

I honestly think that leadership and management training should be started in school … I have never yet seen soft skills training – management skills, personnel skills or soft skills training – I've never seen soft skills training that works successfully with a one hit wonder. You know, go on this course for two weeks and you're a manager. It never works. The only way you can do it is to do it slowly but surely over a long period of time, give people the building blocks, let them go away and use those building blocks, test them on it and then move them on to the next level and move them through their career and target that through their careers. Start when they're a cadet. Start when they're a junior officer. Start when they're in school and actually build up those skills and teach them that you are an officer. That means you're responsible. You are supervising somebody and that's the other thing that we have to get back to. It is a hierarchical management system that we have on the ships.

One noteworthy finding was that the emphasis on training is heavily skewed towards 'seafarer training'. There are only some cursory tokens of training for shore staff. This researcher considers this to be a very unfortunate situation. The merits of certain kinds of training for shore staff were not disputed. The situation is succinctly put by a 22-year-old South Korean Assistant Officer:

> It is company's people (ashore) that do not have knowledge of on board or experienced. So some people don't understand what crew say. Another people ignore saying of crew.

Challenges to Learning

Legal and Socio-Political Limitations

The shipping industry operates in a context that is pervaded by legal norms and consequences. This perhaps is characteristic of twenty-first century life as a whole. The reality of the pervasiveness is not lost on the companies and culpability together with the potential for legal action is always at the fore in considerations of safety measures. Indeed a lot of the safety measures appear to be motivated by a spirit of regulatory compliance and where there are options that could be better for safety these are sometimes stifled because of legal implications. Formal safety measures, procedures and records become even more important since they are basis for claims or protection against claims later.

> Vasilis
>
> You know a lot of such accidents may lead to a very heavy claim, OK. So once somebody will be called to pay, everything has to be supported, well analysed and finding the root causes and what happened in order to have the actual picture of what happened.

Obviously and on the positive side, the prospect of legal action has made the companies more conscientious in their management of safety.

Despite this and the companies' efforts to protect themselves from legal pitfalls with the necessary compliance, there are other areas in which legal implications limits learning in organizations. There is a perception that compliance is driven by the demands of external parties which only want to be 'safe legally' and not because of any commitment to safety per se.

> Tim
>
> People like [names two prominent oil majors] who don't care. They're oil majors, they have a lot of money, they can throw money at things, they don't

care whether it's sufficient or not. They'll throw money, they'll throw manpower, they'll throw lawyers at you. And that's the problem because all that does is prevent you from changing. This is what the lawyers say we must do, therefore we must do it. And that prevents change. That closes the loop. Stops you doing anything. But then there are people out there like us, who are caught in between the two. We need to be efficient, we need to be effective because we need to make money, but we need those people to give us the business. So we have to do some of what they want and we have to educate them and say to them 'look this is a better way of doing it'. And that's what we do.

Alexander speaks of the limitation to the capture and storage of information due to the potential for legal action. As such a selective process (informal and intuitive) is undertaken to decide which information to store in a formal way and which to deal with strictly informally. What this leads to is the limitation of the availability of such information for learning in different times and contexts in the future.

Alexander

Yes [we capture a lot of information informally] because if it is formal – unless of the reporting of near-misses, of facts – beliefs and opinions and let's say, such things, we believe that it's not proper. If it's documented and by names, these things have also let's say, how they call it now by law? I mean you can be taken to the court of law by the opposite. If I make an accusation, which is subjective and cannot be actually justified and we make down in a formal database or record, the other party can claim these records – he has the right to claim the records – and take actions against the people who blamed him and take him to a court of law. That's why many things that are not straight, let's say black and white or not very far from the grey zone, cannot be documented. Unless that we take actions and people on board see these actions, but to be documented is not an easy process. We have been discussing this thing with the lawyers and with the courts and with the P&I clubs for weeks without reaching any end point. And the lawyers just love these cases! They just love it!

As was discussed in Chapter 2, risk has to be considered in light of particular contexts and qualitative values. The perception of safety and risk has significant socio-cultural connotations that ideally should be a constant consideration on the part of the organization in its quest to learn and optimize risk mitigation. However, they are often based on a subjectivity that is difficult to capture in procedures, given the legal implications of such 'evidence'.

Furthermore, the concepts that are acknowledged to be so essential to learning and enhancement of safety are also recognized to be very difficult to measure.

Another challenge relates to the socio-political influences/interests that frustrate optimum safety learning. Whether for purposes of expediency, manipulative control or of survival, the influence of organizational level politics in the determination, taking/running and mitigation of risk is real (Pidgeon, 1997) and has sometimes led to organizational recreancy (Freudenburg, 1993; Pidgeon, 1997; Sagan, 1993, 1994; Vaughan, 1999a). Like it is for all other human institutions, this is characteristic of shipping organizations. The detailed investigation of such processes lies beyond the scope of this research effort. It is only noted that it exists, mediates the construal of risk and can limit enhanced safety learning.

Measurability of Concepts

The contribution of sociological constructs to organizational learning and risk construal is hamstrung by the fact that they are very difficult to measure. The ability to measure is very important to the management of any process, whether regarding safety or quality. In most cases, quality assurance – whether in reference to production or safety – requires that all activities be planned and implemented in a verifiable way and in a continuous cycle of improvement. This planning and implementation cycle should be done vis-à-vis the attainment of stated aims and objectives (Deming, 2000). Measuring is the only way by which such a cycle can be monitored. Unfortunately, though the importance of the constructs is acknowledged, the restriction of immeasurability remains a serious impediment to learning.

For example, in regard to the kind of management training that facilitates learning and openness to diverse perceptions of risk Tim notes that:

Tim

This is the difficulty of actually demonstrating how you do it and is obviously something that the industry as a whole is only just starting to grapple with. The Nautical Institute produced what they considered to be a standard for this kind of leadership and management training taking into account quite broad brush things, culture, communication skills, management skills and it's a huge gap in training that the industry has suffered.

Alexander

But these things are not measurable. If we do it, how well do I measure if it is effective or not? Only by the years we'll have the accidents and incidents will reduce and provided that we give them a safe environment to work; I mean the vessels are well maintained, spares onboard, we give them plenty of time to do their jobs safely, you do not pressure them to do things ... which are there – the situations are already there. These little things, because the big things everybody has it – all the shipping companies. Now how much they believe on those or not is not the question. The big things everybody has it. It's the little things, we believe that, could make ... could [stresses] make the difference. As I said, it's not measurable. It's only in the long run that you will see effectiveness or not. And this is how we try to deal with things.

Despite this acknowledged difficulty in measurement and given that every party recognizes the importance of the soft skills, one would suggest that an increased focus be given to these social constructs in the context of education rather than legislation.

Summary of Chapters 7 and 8

This chapter and the preceding one have discussed the emergent themes in the research that are associated with the theoretical and operational themes of knowledge acquisition, information interpretation and distribution, organizational memory and change/adaptation.

The emergent themes discussed include:

- The dichotomy between near-miss reporting and precursor analysis – noting the bias toward the reactive investigation of incidents as against the analysis of precursors as defined in this work.
- Double-loop/single-loop learning – noting the prevalence of the single-loop kind of learning and the need to augment this with double-loop inquiry.
- Cultural interaction/team psychological safety/leader inclusiveness – noting the role culture plays in team psychological safety and other team dynamics and the more significant role played by strict hierarchy and rank differentiation.
- Blame/accountability – noting the effect of blame on learning and difficulties in striking a balance between blame and responsibility/accountability associated with competence in high-risk industries.
- Worker engagement – noting the important role worker engagement plays in learning, and differentiating between direct safety approaches and more indirect but holistic approaches, as well as between opportunities for worker engagement and motivation for worker engagement. Further noting that such holistic approaches/measures include crew retention, salary, social relations and amenities on board, team dynamics, contracts and crewing links with owners, professional development and social and job security and welfare.
- Accident causation philosophy – noting how the kind of philosophy held significantly affects risk construal and learning and the tendency in the industry to rely on basic (but practical) sequential modelling of the accident causation process and industry's hesitance to delve into contemporary accident causation debates in academia.
- Resource availability/utilization, entropy and drift – noting how shipping organizations are affected by drift, natural entropy and how resource

availability and utilization affect learning. Further noting how resource availability/utilization is linked to the pervasive view (in both industry and academia) that productivity goals are irreconcilable with safety goals and how some companies show that this is a flawed perspective.

- Finally noting and discussing some challenges to learning in legal and socio-political limitations and difficulties in measurability of concepts.

Chapter 9

Research Conclusions, Implications and Recommendations

Everything at its proper time and with proper attention to existing conditions. There is no universally applicable rule, no magic wand, that we can apply to every situation and to all the structures we find in the real world. Our job is to think of, and then do, the right things at the right times and in the right way.

Dörner[1]

Purpose and Outline

Practitioners and managers in high-risk industries are often averse to the findings of scientific communities (for an extreme example of this see McHugh, 2007, p. 18). These findings seem to them as vague, ivory-tower-closeted views and even where such findings are made relevant through legislation, they remain – in the mind of some – manifestations of an outside world, one that is unreal, impractical and just muddies the waters of 'getting the job done'. Academia must seek to bridge this practice-research gap. This chapter seeks, in conclusion to this work, to discuss some practical implications of the research carried out in various maritime contexts and to give some recommendations for the industry and other stakeholders. The full original findings of the study are presented in Chapter 5. The findings indicated in this chapter are those on which recommendations are derived.

Research Conclusions

The major research focus for this work was to explore how shipping companies as organizations learn from, filter and give credence/acceptability to differing risk perceptions and how this influences the work culture with special regard to group/ team dynamics and individual motivation. The study began with the exploration of constructs that could be hypothesized to be associated with the global research question and an examination of the literature for theoretical underpinnings for these constructs in the fields of organizational theory and learning, risk management and team/organization dynamics and in the specific context of the maritime industry.

---

1   From *The logic of failure: Recognizing and avoiding error in complex situations* (1996). Translated by R. Kimber and R. Kimber and published by Metropolitan Books in New York.

This review of the literature not only pointed out long-standing discussions in the areas of risk and learning but also generated definitions to be utilized in the work. It also afforded the opportunity to ground the subsequent research on justifiable theoretical bases and to seek to address the gaps in the research especially as it relates to the maritime industry. Among other things, a basis was found in the literature for the view that risk can be seen as a subjective and hierarchical epistemology and as an ontological objective reality. The subjectivity inherent in risk cognition and management, make it vital that processes of risk construal and communication be given significant importance in studies in high-risk industries. The study also found merit in the framework of social amplification/attenuation of risk and introduced the neologism 'amplenuation' to avoid the bias in the literature towards emphasizing *amplification* over *attenuation* of risk signals. A number of constructs were defined and found relevant to the work. They included team psychological safety (TPS), leader inclusiveness (LI), worker engagement (WE) and organizational learning (OL). The theoretical review also created a basis for the generation of a number of theoretical and operational themes that could be operationalized in the quest to answer the global question. These included *three theoretical themes – Individual/organizational interaction in learning, levels and types of learning* and *the role of knowledge structures*; and *two operational themes – organizational learning processes* (knowledge acquisition, information interpretation pre and post distribution, information distribution, organizational memory) and *organizational learning outcomes* (change/adaptation).

The constructs and themes then informed the raising of specific questions and the methodology to be used in the research:

- What are the relationships between team psychological safety, leader inclusiveness, worker engagement and organizational learning as regards the context of shipping companies and ship officers?
- How do the constructs of team psychological safety, leader inclusiveness and worker engagement, predict organizational learning?
- Are the predictive potentials of team psychological safety, leader inclusiveness and worker engagement for organizational learning confounded by variables such as rank, age, nationality and time in company?
- What variables influence worker engagement in ship officers as regards safety?
- What are the processes that facilitate or limit organizational learning in shipping companies?
- What factors influence credence-giving as regards risk information from ship officers?

To address these questions a mixed-methods (multi-step) approach was used. It involved the use of a focus group discussion, a survey/questionnaire and field visits for in-depth interviews and document/software perusal. In recognition of the individual-organizational interaction in learning processes, a structuration theory approach was used in focusing on the recurrent practices of individuals in an organizational setting as evidenced by their responses to the questionnaire, in the interviews and from the shipping companies' procedures/documents/software.

The research found a number of significant positive correlations between team psychological safety, worker engagement, leader inclusiveness and organizational learning (knowledge acquisition, information interpretation and distribution, organizational memory and change/adaptation). Among other things it was found that team psychological safety and worker engagement in particular are significant predictors of organizational learning. Also found were other emergent themes that influence organizational learning with respect to safety: predominantly resource availability and utilization, productivity versus safety, entropy, drift and migration and accident causation philosophy, precursor analysis and blame/accountability.

The overall conceptual outline of the research findings is shown in Figure 9.1. The figure indicates:

- Links between knowledge acquisition (congenital, experiential, vicarious and inferential) /interpretation/distribution and external/internal influences as per open systems theory.
- The synergies between individual/team/organization as per structuration theory.
- The mediating role of individual subjectivity/perception in the amplenuation of risk signals.
- The mediating role of worker engagement, team psychological safety and leader inclusiveness regarding functional input in organizational safety strategy and action.
- Factors that influence capacity to act at different stages of learning.
- The mediating role of socio-cultural norms.
- Organizational memory and communication.
- Organizational change/adaptation and the progressive development of organizational culture.

This organizational learning and risk construal process is cyclical. Some form of learning is going on all the time. Where organizations recognize the factors at play, as depicted in Figure 9.1, efforts can be made to optimize the process(es) for productive learning.

Achieving quality (as evidenced in consistent/continual high safety standards and performance, environmental protection and sustainability and secure maritime transport) is based on the application of good management practices with regards to the amplenuation of risk signals as well as intentional and structured processes of organizational learning. These practices (and attitudes) will necessarily include management commitment and the greater consideration of all risk perceptions – particularly from the operational (sharp) end – than appears to exist now. It may be argued that the hierarchical credence-giving or acceptability of risk perceptions is a de facto situation – and rightly so. What is required is that it is premised

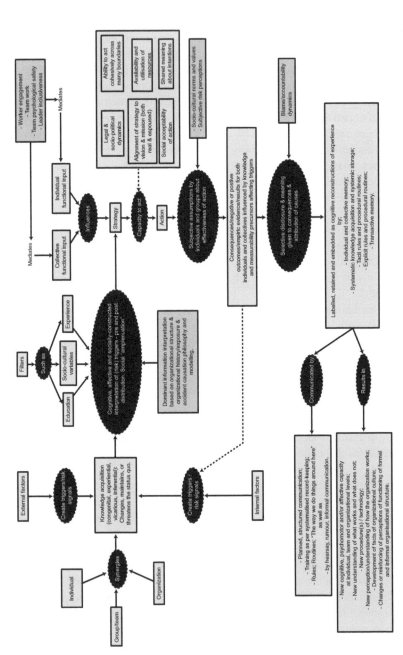

Figure 9.1 Conceptual representation of processes associated with risk construal and organizational learning

on an awareness and appreciation of, and unobstructed access to different risk perceptions.

It is obvious that there are attempts to learn to various degrees in the industry. However, the learning in shipping organizations is essentially one of single-loop learning, being characterized more by adaptive learning than by higher-order reflective learning (Pawlowsky, 2001, p. 75). This single-loop kind of learning appears to be partly the result of an undue reliance on law for social control in the maritime industry. Safety action in the industry is very much motivated by legislation. The consequences of this are deemed not to be in the best interest of safety in the industry. Both safety science and the law have ends in non-recurrence of accidents. Maritime organizations, however, are very much aware of the purported conflict between ideal safety science objectives in accident investigation (personal, organizational and industrial learning) and the perceived legal objectives (of liability, penalty, sanctions and compensation) which create an atmosphere for blame. The ideal can only be reached where maritime organizations show – clearly and consistently – that they are driven in their operational choices regarding safety as much by organizational learning as they are by profit margins and externally imposed compliance regimes. Where the global community at large has the perception that profitability drives the maritime industry at the expense of safety, regulatory measures and legal sanction and penalties will continue to increase, thereby creating a vicious cycle of blame, error-hiding and potential for rapid deterioration of safety. As acknowledged by some respondents in the companies visited – which are higher than average in being safety-focused – there is substantial satisficing in shipping with definitions of the acceptability of risk that is motivated by commercial interests.

The reliance on law is also manifested in the industry's heavy reliance on the use of negative consequences of bad behaviour in seeking to achieve proper operational behaviour. This is evidenced by the proliferation of various legal instruments which seek to 'punish' offenders. 'Rewards' in this context, exist only in the avoidance of penalties.

It is recognized as per this research and the literature that good regulatory regimes do lead to safety benefits. Over the years the global threshold of safety has been raised. Legislation like the ISM has made explicit and perhaps more measurable what (for the more safety-conscious companies) used to be implicit/ intuitive ways of approaching safety. For shipping organizations characterized by recreancy, ISM forced the creation of verifiable safety management systems where previously these might have been non-existent. Legal imperatives have also helped give voice to the health, safety and quality elements of shipping organizations – elements which together with the issues they tend to raise – were often lost in the all-pervasive and increasing dominance of the commercialization of shipping that was threatening to drown out professionalism and safe behaviour. Generally speaking therefore – and despite the acknowledged existence of many challenges in the implementation and enforcement of various legal instruments – the legislative approach has been relatively successful.

However, regulatory law cannot oversee everything even when it is minutely and infinitesimally prescriptive. Regulations breed more regulations and with it a compliance attitude that focuses inordinately on passing audits (and the buck) because of liability issues and not because of a fundamental commitment to safety. A skewed emphasis on the regulatory regime is like trying to pick up a nail and hammer it into a piece of wood with one hand. With some dexterity, it may be possible, but it would be best if at least two other hands help to hold the nail and the piece of wood in place. Legislative instruments alone cannot meet the ambitious goal of bringing about what is essentially social change in safety and the maintenance of a positively dynamic focus on safety in the maritime industry. There are three key approaches to effecting social change. One is by legislation, another by norms (via, for example, education and role-modelling) and a third by markets (Schuck, 2000). Social norms (via education at 'school' and in 'industry') and the market (for example, as observed in the influence of TMSA) can help law achieve the ends it desires. Of course this point of view does not naively assume that with 'better' education (and understanding of the socio-psychological dynamics of risk construal and organizational learning) and markets as the 'other hands' to help 'hammer the nail', all problems of safety and risk governance will disappear. There are too many individual, corporate and institutional interests for that 'miracle' to happen. The possibility nevertheless does exist for an improvement which will necessarily be correlated to the will and the resources the global community (and specifically the maritime industry) are prepared to commit to this. It is the view of this author based on the research, that a significant focus on education – not only regarding procedures, but more importantly the fundamental principles of risk (including the sociology of risk), safety management systems and values – will not be misplaced. Despite the substantial benefits of legislation, it is taking the global industry some time to wean itself off the prescriptive model of legislation and adapt fundamentally to the 'duty of care' that, for example, the ISM Code calls for. To achieve this, the maritime industry as a whole must become more sensitized to what a few are focused on: the need for education/training for all actors in new competences with new educational paradigms that are based on new understandings of risk, safety and the management of both.

This work puts forward the notion that currently the market dynamics of economics and rewards for meeting schedules and deadlines strongly favour the violation of what is often perceived as banal safety regulations. Without the positive articulation and overt insistence on a culture of learning and safety before all else, seafarers will continue to make choices based on perceptions of shore-based pressures for economic expediency. It does not matter that the company falls back on 'we never told him to sail' positions. Even clear statements in company manuals calling for the respect of a master's judgement, will not take away the potential for a latent awareness of the primacy of economics over safety. In a very commercially competitive world, the de facto message that in many cases appears to get across to ship officers is that sometimes (whether expressed overtly or implied by other company priorities) cutting safety corners pays. This can only be countered, not by passive acquiescence to current legislative requirements for formal statements in management documents, but by vigorous affirmation of the primacy of the preservation of life, property and the environment over economics.

It would appear that it is not enough simply to state in formal documents that seafarers should not hesitate to point out their perceptions of risk as part of engagement and learning. The ISM Code, for example, specifically makes reference to a master's authority in determining and proceeding with what (s)he perceives to be the best course of action. Indeed, as far back as the days of the SS *Titanic*, more stringent measures existed. The White Star Line (owners of SS *Titanic*) required that their masters acknowledge in writing their understanding and affirmation of this authority ('Report on the loss of the S.S. *Titanic*,' 1990, p. 24). This is a standard beyond that required by the ISM code. Nevertheless, 'time pressure' may have contributed to the *Titanic* proceeding at an unsafe speed (22 knots) in areas known to have ice. Something else is communicated to ship officers beyond the normal sets of documents that are read and signed. As in the case of the *Green Lily* accident (mentioned in Chapter 1), and as evidenced in all of the companies visited for this research, many ship officers receive no overt external pressure to reduce safety margins, but may make decisions to do so based on more latent pressures arising from the tacit assumptions that are reified in the shipping organization and the shipping industry as a whole.

Indisputably, safety learning and risk construal – in the sense in which it is defined in this study – is a consideration for members of the maritime industry (generalizing the results of this survey). Teamwork and organizational acceptance of its importance have come a long way in the maritime industry. However, it is also clear that there are varying degrees to which importance is attached to it by way of what is expected, what is valued and rewarded/punished and at what level this importance is evident. The research suggests that shipping companies are more prone to attach importance to teamwork when it is related to teams on board. The necessity of risk perception sharing between ship and shore and the necessary presence of formal systems of knowledge acquisition, information interpretation and distribution in this area is neither explicit nor ideal, despite the legislative requirements of ISM. Perceptions do matter and there seems to be empirically supported evidence that an unduly high percentage of ship officers have the perception that shore management does not regard seafarers' (in this case ship officers) risk input as much as the seafarers would deem fit. Granted that this acknowledgement of one's epistemology of risk is correlated to worker engagement and input into team learning and thence by extension into organization learning, it can be concluded that there is room for enhanced organizational learning in the maritime industry.

The research found substantial correlations between the concepts of team psychological safety, leader inclusiveness, worker engagement and organizational learning. Organizational learning is fundamental to safety optimization and is informed by a risk construal process that appreciates the input of all levels of the organization. Optimum knowledge acquisition, as a key part of knowledge management, requires that organizations create cultures, platforms and systems that capture the collective expertise of their employees and facilitates in non-latent/ non-tacit ways, the collaborative sharing of knowledge embedded in individuals and teams (Gordon, 2002). This requires a management style and culture which acknowledges the increased empowerment of seafarers at all levels as knowledge workers and as significant contributors to the safety knowledge pool. This culture, while present in some shipping companies is not prevalent in the industry as a whole.

The research shows a significant reactive culture that is dependent on a lot of near-miss reporting both as a means of worker engagement in the safety process and as a way of learning lessons without the costs of accidents. This is a paradigm that is heavily reliant on hindsight in keeping with most of human life which is present monitoring in light of the past – an ongoing review of lessons in hindsight. However, relying on hindsight – the natural bias of organizations – has obvious limitations as has been pointed out in this work. On the other hand, perfect foresight is not possible with humans and the systems they create. Contemporary corporate risk management does not engage in 'prophecy' – at least not officially. Prophecy is deterministic: this and/or that *is* going to happen. Humans can do no better than predict.[2] Predictiveness in this context is conditional and should be based on *hindsight, insight* and *foresight*, limited as they are by the human condition. To make the best of foresight, organizations must incorporate insight. The research suggests that insight is gained into all aspects of a diverse safety/ risk environment when there is access to all sources of information and where there is an environment that espouses *and* supports the articulation of diverse perceptions (of individuals, teams and groups) of risk. Organizational behaviour that is significantly biased towards hindsight is risky. It is similar to driving with your attention only on the rear view mirror. It makes a culture of blame thrive and leaves the organization vulnerable to threats resulting from the dynamism inherent in the emergence of diverse contemporary internal and external environmental factors. As Dörner notes:

> The dynamics inherent in systems make it important to understand developmental tendencies [trends]. We cannot content ourselves with observing and analysing situations at any single moment [feeding into single-loop learning] but must instead try to determine where the whole system is heading over time ['drift' a la Rasmussen]. For many people [and organizations] this proves to be an extremely difficult task. (Dörner, 1996, p. 40)

The difficulty Dörner alludes to can be ameliorated by better organizational risk communication and construal which could evolve through education. As Maslow (1966, p. 15) noted 'it is tempting, if the only tool you have is a hammer, to treat

---

2   To indicate what one thinks could or will happen in the future based on present conditions or observations, on past experience or on scientific extrapolations.

everything as if it were a nail'. If ship officers and managers are trained, educated or equipped with particular risk paradigms, they will look for the 'nails' that suit those models. If ship officers and ship management have training that focuses on hindsight-biased or reactive paradigms of safety, they will tend to see all risk problems in that light and act in keeping with that view. A potential solution may lie in enhanced managerial and operational education, possibly in the context of the trend in METI towards the awarding of degrees. Furthermore an increased focus on ship officer competence beyond technical ship operation skills and with more emphasis on education in risk and learning theory, will lead to more credence given to ship officers by other significant constituencies in the industry. This kind of education will also improve their job prospects once they leave the ships as active seafarers, thus making a seagoing career an attractive option for young people.

Practical Implications and Recommendations

A number of research findings were arrived at, at the completion of this study. Recommendations are based on these findings.

Finding: Risk Definitions

The pure objective definition of risk as probability multiplied by consequence is limiting. It tends to ignore very real risk issues which are more subjective and on which risk policy depends. The purely objective notion of risk does not subsume the elements of the sociology of risk (for example, perceptions) as is the case taken for example in Renn's work on 'risk governance' (Renn, 2005). The standard risk management process does not address sufficiently the issues of the acceptability of risk, risk policy setting and decision-making.

Recommendation It is recommended that the industry as a whole should be more open to the sociological parts and connotations of risk and expand the scope of dealing with risk from management to governance. Risk governance includes in its discourse sociological factors including socio-cultural, legal and policy dimensions that impact on risk choices, acceptability and tolerance, much more than do current risk management approaches.

Finding: Worker Engagement

- Not only is worker engagement positively correlated with perceptions of organizational learning and team psychological safety, the two are significant predictors of organizational learning.
- Worker engagement is influenced not only by direct solicitations to ship officers to be involved and participate in specific safety-related

programmes, but also by more holistic factors that, prima facie, may not appear to be safety-related.

- Worker engagement levels for management level officers are moderately higher than those for operational level officers.
- Based on manning/labour legislation from different countries, sometimes crew from different nations are paid different wages for same work done. While these actions may be administratively justified, they nevertheless generate perceptions of inequity and adversely affect motivation and engagement.

Recommendation

- Shipping companies should give voice and action to policies that encourage the factors that increase worker engagement.
- The use of safety circles, as direct solicitations for worker engagement in safety and especially between ship and shore, must be encouraged.
- Post-task discussions by team members should have as much emphasis as pre-task discussion.
- High retention rates are critical and show that companies that do not use this as a key performance indicator may be compromising advances in the achievement of a safety culture. High retention rates help to engender an organizational acculturation process that limits the role of nationality/personality differences in inhibiting a safety paradigm. In the tanker industry there is a requirement (per TMSA) that this be a key performance indicator. It is strongly recommended that all sectors of shipping consider the merit of making this a requirement.
- Other holistic factors that contribute to the achievement of worker engagement include salary, professional development, amenities on board, social security and welfare, contract links to owners and social relations on board. When considering worker engagement in safety, companies should appreciate that, further to the direct practices that engage crew in safety, these more holistic factors create enhanced identification with organizational goals including safety goals.
- Companies should take note and reflect on the effect on teams – and by extension learning and safety – of different national legislation for multinational crews, which allows for disparities in wages (for same work roles) based on nationality and the connotations of racism it creates. Companies must clarify the reasons for differential salary scales to crew or in much better contexts seek to avoid the practice.

Finding: Team Psychological Safety and Leader Inclusiveness

- Team psychological safety is positively correlated with perceptions of organizational learning, worker engagement and with leader inclusiveness.

- Team psychological safety is moderately higher for OECD officers than for non-OECD officers and significantly higher for management level officers than for operational level officers.
- Leader inclusiveness is positively correlated with perceptions of organizational learning and with worker engagement.

Recommendation Shipping companies will benefit from a climate in which all – especially operational level staff and non-OECD officers – are uninhibited in their sharing of risk perceptions. The purported gravity of nationality differences is not supported fully by the research, but the difference in team psychological safety as a result of differences in rank is noteworthy. It behoves the shipping industry and the educational institutions that support it to make encouragement, the solicitation and acknowledgment of inquiry-led contributions an index of competence and to educate, especially management level staff, to view this as a competence issue.

Education should also emphasize the negotiation of reality – the communication skills that transcend nationality and that are not fundamentally stereotypical. Hofstede prominently states on his website[3] that 'culture is more often a source of conflict than of synergy. Cultural differences are a nuisance at best and often a disaster'. While acknowledging the extensive work that must have informed this conclusion and perhaps its applicability as a management view in other contexts, this author disagrees that this is a view that should be taken by organizations especially in the arena of global ship operation and management. On the contrary, diversity (including cultural diversity) may be seen and utilized as an asset. Differences and conflicts in perceptions and behaviour, if accommodated and negotiated appropriately, may actually be an asset for any organization and not a nuisance and certainly not a disaster (Berthoin Antal and Friedman, 2005; Graham, 1995, pp. 19–20 as cited by Child and Heavens, 2001; Rothman and Friedman, 2001; Sjöberg, 2006). A society's metaphors are compelling and have a significant effect on that society's policies, strategies, mental maps and behaviours (Morgan, 2006). Picturing organizations as an orchestra – in turn a classical orchestra and then a jazz orchestra – gives a powerful picture of how organizations can have finely tuned leadership that draws out the merits of diversity and on occasion allows the free-willed improvization of expert and competent players and learners to make better 'organizational music'.

Finding: Credence Giving

The majority of officers (66 per cent) think that their safety-related opinions are sometimes or always considered by shore-based management or that they have power to change a management decision. However, a significant 34 per cent feel that their opinions are rarely considered, that they have no influence on management or were not sure.

---

3   http://www.geert-hofstede.com retrieved 10 June 2009.

Recommendation Measures that incorporate the risk perception of ship officers should be made explicit. Action that results from such input should be acknowledged. This motivates and leads to increased engagement.

Finding: Congenital Learning

Congenital learning has a significant effect on the 'knowledge corridor' that is created for subsequent learning in an organization and also has implications for organizational entities that have their origins in specific companies.

Recommendation Safety management in shipping companies must recognize the effects of congenital learning and inquire as to the effectiveness and merits of long-standing tradition and premises for the extant safety philosophy.

Finding: Temporal Demarcation in Safety Data

The shipping companies have 'calendar-based demarcations' or 'organizational breakpoints' for safety information distribution, trending, analysis and discussion. This may introduce salient points that are not necessarily relevant or important to risk management. Salient features in a stream of information affect managers' perceptions of causality, relevance or importance. Segmentation of safety information may influence how risk and solutions are framed and how resources are allocated.

Recommendation There is an acknowledged need for calendar-based demarcations in safety operation, reporting and verification. Despite this, managers must think of safety as a continuum so that the salience of break points does not distort the relevance/importance of prior safety information.

Finding: Top-Down Appraisal of Risk Associated Competence

Information acquisition about risk associated with crew competence is primarily derived from top shipboard management concerning lower-level crew.

Recommendation There must be room for lower operational crew to give formal appraisals of higher-level management performance or competence, in a manner that does not compromise team dynamics.

Finding: Ship Visits

Some companies place very highly the visiting of ships by very senior management. Indeed it is a requirement for tanker-owning companies subscribing to TMSA. However, there is an obvious assumption that people who visit the ships are equipped to appreciate the differences of context, an understanding of what constitutes normality in safe ship operation as well as the learning processes involved and a double-loop inquiry mindset.

Recommendation   While the assumption may be valid in specific cases, it should not be over-generalized. This is another area in which more education in risk cognition and the examination of approaches to double-loop inquiry would be beneficial. Such training as has been suggested can be offered in the context of degree programmes so that future ship managers and crew have a perspective on risk that is more in-depth than they currently have.

Finding: Vicarious Learning

The shipping companies place value on vicarious learning. However, this is not as systematized as learning from the organizations' own experiences which is covered by many safety management systems. Neither is the issue of contextualization of vicarious learning systematized.

Recommendation   The existing safety management systems should be made to incorporate more vicarious learning with the necessary contextualization reflected.

Finding: External Databases

At the level of shipping companies, the IMO database GISIS and FSI circulars are not an immediate point of reference for lessons learnt. The IMO database is not publicized enough and its merits not sufficiently apprehended by companies and their crew. The databases are also almost exclusively and necessarily computer- and web-based. Therefore the (acknowledged) severe limitation of internet services for crew on board ship is a very significant setback for direct crew reporting. The research found that the limitation of internet service is increasingly proving to be an issue/problem with seafarers (especially those from more developed countries) not just regarding reporting but as a social amenity onboard.

Recommendation   Companies must see the merit (as some already have) in good internet access for ship crew and commit resources to this. Apart from improving reporting, it will serve to create better social links between seafarers and others ashore (whether private or organizational) and thereby increase worker engagement (via the social amenities on board link). There is a need for more prominence of a global, advertised, accessible database for the sharing of learning experiences across industry – vicarious learning – and in MET. The IMO has a key resource – in the World Maritime University – to operate, maintain and analyse data from an industry wide database along the lines of the ASRS found in aviation and CCPS in the chemical industry or Sweden's national level INSJÖ for the maritime industry. Using the university as a non-profit, non-biased, world centre for this (with funding generated possibly from industry) would create an internationally recognized and impartial collection centre for key safety data. Like the ASRS, the necessary scope and legal waivers will have to be in place to attract the relevant data and to perform the necessary analyses. The output of analyses on such data would be invaluable for the industry, the academic advancement of the subject area and the global community at large. Such a database would serve a purpose greater than the periodic extraction of incidents for reading for lessons learnt, and importantly will devote qualified human and technological resources to the careful analysis of data for underlining patterns for research.

Findings: Experiential Learning and Near-Miss Reporting

Knowledge acquisition in the context of experiential learning is very biased towards the reporting of near-misses. The quality of near-miss reporting is not such as is supportive of optimum proactive management of risk.

Recommendation   Precursors should be viewed more in the light of stable conditions that herald accidents and near-misses. Reporting should be focused on precursors as well as near-misses and accidents. This more proactive view can be facilitated by education of ship officers and shore-based management that is increasingly directed at a fundamental understanding of the role of precursors in accident causation and such latent themes as the sociology of risk. The need to incorporate risk perceptions in the cognition and control of risk should be emphasized in both industry and educational settings. Within the current trend for degree-awarding for ship officers, it should be possible for a more theoretical curriculum of risk to be included in the ship officer's education.

Finding: Trend Analysis

In the information interpretation phase of organizational learning, trend analysis, though mentioned by many of the interviewees, was still very much an accounting approach, what Woods and Cook call 'counting failure' (Woods and Cook, 2001) and not very focused on deeper level analysis nor sufficiently on precursor analysis in the anticipation of risk. The full potential of information technology in the interpretation of data is yet to be exploited by shipping companies.

Recommendation   The undue emphasis on learning from mistakes is limiting. Learning from mistakes should never be discounted and serves as a necessary primary stage for learning. However, more emphasis should be placed on learning based on precursor conditions – technical and social – and their communication. This draws once again attention to the importance of feedback loops and make it equally important that companies create the environment – and educational institutions, the 'competence' – that allow for speaking up even regarding the often non-objective factors that are essentially precursors. Companies should invest in systems that support the deeper analysis that departs from the accounting approach and consider the employment of key personnel (as BP was said to have done) in analysis of data (which should include precursor information as defined in this work) to inquire into patterns of causation that are not immediately obvious, for example as relates to resource allocation and sociological determinants of risk. A few of the companies visited recognize this and are already working in this direction. It is highly recommended that the industry at large emulate this as an appropriate benchmark.

Finding: Voice and Silence in Leaders

The research found cases where key policy-setting or policy-influencing personnel articulated positions contrary to the existing organizational policy (theory-in-action), and yet showed no previous action or inclination for future action to change the status quo. This reflects the restriction of voice even in leaders. The literature often views the issue of voice and silence in organizations as an issue with subordinate communication with superiors (upward communication). The finding shows that issues of voice and silence are not restricted to upward communication but also affect lateral communication.

Recommendation   Leaders' need to voice perception must be acknowledged by the research literature, ship management, educational institutions and all leaders. Education should highlight the tendency for the imbalance that ignores voice and silence in leadership and train accordingly.

Finding: Transactive Memory

For good and practical reasons, training ship teams (off-ship) together for the purpose of improving transactive memory is difficult and perhaps unattainable.

Recommendation   In recognition of the importance and relevance of transactive memory, but as a substitute for training together, companies should emphasize and work to attain uniformity in optimum and company-specific education/training, whether this training is provided by traditional METI, company-owned institutions or other third parties. Safety managers must be aware of the role that transactive memory can play in organizational memory and learning. In the dry trades, where high retention rates may not be a required KPI, it is strongly recommended that ship management nevertheless make it a necessary focus. High retention rates, increases members' familiarity with other members' roles and memories in specific company settings and have benefits in better risk communication and management.

Finding: Ship–shore Communication

Semantic differences in language and differences in risk perspectives inhibit risk communication between ship and shore.

Recommendation   More people with relevant seafaring backgrounds should be employed in shore management. This gives significant relevance to the trend in educational institutions in running degree programmes that allow for a career path beyond operations on board ship for seafarers. While this course of action may be restricted by the current perceived global shortage of ship officers, it nevertheless should be what companies aim for and are prepared to commit resources to.

Findings: Hierarchical Structures

Management structures on board remain significantly hierarchical and there is resistance to the concept of flatter hierarchies. However, the strict hierarchy and current understandings of responsibility and rank, detract from optimum team psychological safety. The research suggests that the strict hierarchy (in the way it is operationalized today), constitutes more of a challenge to increased team psychological safety than do nationality differences. There is, however, more awareness and emphasis placed on nationality issues by the companies.

Recommendations   The nature of work teams (in all fields of endeavour) is evolving from static descriptions of individuals working in functional structures to more 'dynamic, emergent, and adaptive entities embedded in a multilevel (individual, team, organizational) system … dynamic systems that exist in a context, develop as members interact over time, and evolve and adapt as situational demands unfold' (Kozlowski and Ilgen, 2006, p. 78). While it may not be possible in the short-to-medium term to drastically change management structures on board ship, it is recommended that – via education – there is a paradigm shift in what it means to be a team member/leader and that as an index of competence, there is an association of teamwork and the emphatic solicitation of diverse views into the risk cognition and management process. This should be attended to with at least as much focus as that applied to nationality issues – if not more.

Finding: Research-Practice Gap

Industry does not appear to view the discussions in academia regarding for example accident causation and modelling as practical. Intuitive modelling is done with no recourse to contemporary critiques in the literature. Industry in general and the maritime industry in particular place an emphasis on simplicity and practicality. These are not necessarily characteristic of academic debates. It has been noted that daily decision-making by individuals working in complex organizations is 'shaped more by power structures, ingrained routines, and established resource configurations than by current scientific findings' (Rosenheck, 2001).

Recommendation   Academia must find a way to bridge this research-practice gap by, for example, the development of suitable educational and training courses and the progressive development of practical tools based on research findings. The presence of a tool that helps in the auditing of shipping organizations as regards organizational learning and its influencing factors, provides 'a basis for moving from a traditional bureaucracy [single-loop learning systems] to a learning organization, thereby improving the way in which organizations function' (Pace, 2002, p. 458). Perhaps what is even more important is not so much the existence of tools, but a paradigm shift in mental models of management and the resulting management styles. Auditable tools are being used as protection from liability and do not seem to address fundamental system functioning. Intangibles and things difficult to measure are not prioritized. Deming (as cited by Senge, 2006, p. xiv) notes that 'you can measure only three per cent of what matters'. Accordingly, this book, while recommending the progressive development of tools to bridge the research-practice gap, notes the even more important need for education that is geared at all levels towards fundamental understanding of the principles on which risk construal and governance as well as organizational learning are based. The most pressing requirement is the consideration of organizational learning as an active management process.

Finding: Productivity/Safety

Both in academia and in industry, there is a tendency to view productivity and safety goals as incompatible and irreconcilable. This leads to questionable practices of resource allocation and utilization that compromise safety in the name of productive efficiency.

Recommendations   There is evidence from this research suggesting that this pervasive view is flawed. Enhanced safety goals can and should be reconcilable with higher levels of productivity. The holistic factors that improve worker engagement and thus organizational learning are not restricted to safety. Engaged workers identify with all goals of the organization (including safety) and this betters productivity. It is recommended that the industry (as a whole) does what a few are doing and break out of the restrictive paradigm that suggests that resources committed to safety are 'lost' to productivity.

Finding: Training

There is an increasing burden of training being placed on the shipping companies. Those who can afford it are even establishing training centres that function almost like the traditional METI. This would suggest a deficiency in the work of traditional METI or could be due to a lack of communication between industry and METI (academia). Generally speaking, traditional METI appear to be completely out of the loop when it comes to current and dynamic risk information distribution.

Recommendation METI (generally speaking) should work to enhance communication between themselves and industry so that their curricula can be relevant to the needs of the companies beyond STCW. Companies (and METI) will benefit from formal communication links that have 'feed-in' for sending recommendations for curricula upgrades based on lessons learnt so that METI will be relevant to industry beyond legislative compliance and the investment in MET infrastructure will be limited for industry.

Finding: Shore Staff Training

The emphasis on training is heavily skewed towards 'seafarer training'. The research suggests that in the area of communication, safety, risk construal and shipboard operation, shore office personnel would also benefit from enhanced training.

Recommendation   It is recommended that both industry and MET recognize the need to train shore personnel (in a professional development or refresher context) as much as they do seafarers.

Findings: Motivation for Safety Action

The main motivation for safety action seems to be legal and/or commercial imperatives. Such actions in most cases cover things that are easily auditable. The research shows how less tangible constructs may play a role in the generation and maintenance of a safety culture. The companies, however, seldom attempt to formally measure these constructs, relying on intuition and gut feeling to assess this.

Recommendation Independently of the legally required audits, shipping companies will benefit from periodic attitude surveys done by independent consultants (for example, universities/researchers) to give an indication of company standing with respect to constructs like team psychological safety, worker engagement and leader inclusiveness. Such surveys, as noted by Baker, King, McDonald and Horbar (writing in the context of the medical industry 2003, p. 419) 'can provide useful information on team and organizational issues that impede improvement [and] ... promote discussion of team issues and assist in identifying changes in policies, communications, and interactions that promote more effective team behaviors'. They will be mutually beneficial to industry and research in academia. These kinds of 'audits' should have no legal consequence, be not so frequent as to lead to questionnaire-fatigue and have their findings publicized (at least within the company) and most importantly, acted on. Furthermore, the difficulty in measuring/auditing some of the constructs and themes that affect organizational learning (in the context of industry), puts a greater emphasis on and highlights the place of education (in the context of schools) in building this learning culture as a key part of competence.

It can be seen that many of the recommendations revolve around education. This author considers this the primary recommendation of the book – enhanced non-technical education that places importance on learning and risk construal processes.

Theoretical Implications and Recommendations

Social Exchange Theory

Discussions about social exchange theory have centred on theories of operant conditioning (Homans, 1958), views on productive exchange (Emerson, 1962) and micro economics/rational choice views (Blau, 1964). Much more recently, there were the additions of the emotional/affective component of exchange (Lawler, 2001; Lawler and Thye, 1999) and Cook's consideration of cultural context (Cook, 2000).

To this evolutionary trend, this research adds the consideration of the multiplicity of exchange contexts (between an individual and different entities) and the resulting choices based on relative value with possible variables of proximity in time and space. There is merit in the view of social exchange theory as applicable to more than dyads and that value-based exchange relations can be exist between individuals and macro-social entities such as organizations. In this case it is worth noting that there may be multiple such exchanges going on at any one time. Take the hypothetical case of an exchange scenario between a subordinate and his/her superior which is not rewarding. According to the theory, this will lead to the reduction of commitment or avoidance of the action that any one party considers valuable to the other. At the same time it is plausible that this action may be considered valuable by another entity, for example other dominant personalities in the organization or the anthropomorphized organization, and that in that exchange relationship with the subordinate, there may be mutual benefit and a reinforcement of the particular action. Will the action then be carried out? There is some evidence from the research that such an action would indeed be carried out. In other words where an organizational culture is relatively enforcing of a productive exchange between a subordinate and the organization as such, intermediary supervisors (like masters) may have limited effect on the relationship between subordinate and organization. It would seem that in this discussion one will have to introduce such variables as perception of relative values and their proximity in time and space. The question that is pertinent is whether organizational factors in the shipping industry (generally speaking) are of enough value to offset the potential individual actions of superiors that may limit the reporting behaviour of subordinate officers assuming the subordinates do not find the social exchange with their superiors productive and valuable. Social exchange theory applied to the maritime industry would suggest that not addressing some of the fundamental issues addressed in this research (especially those related to worker engagement and the perceived organizational support context) may be detrimental to the long-term sustainability of the industry as a competitive attraction for key human resource. If organizational dynamics of the industry as a whole are not perceived to be supportive of the individual, the tendency will be to abandon careers at sea as soon as some other exchange context offers equivalent or greater value. The costs of a life at sea would include – in a world increasingly characterized by uninhibited workplace communication – being unable to feel part of a team because of the adherence to strict hierarchical structures and a pervasive and invasive blame culture.

The Social Amplification of Risk Framework

SARF is most prominently addressed in existing literature at the macro-level, examining dynamics of risk amplification and with such factors as the media playing a dominant role. This research shows how similar dynamics exist at a micro-level (team and organizational settings) with more emphasis on the attenuation of risk signals. This side of SARF, though not emphasized in the literature, is shown to be relevant to risk cognition, management and governance. As shown in Figure 9.2, similar dynamics as in the macro-level SARF attenuate or amplify risk signals with the ripple effects on shipping companies and their teams/employees, that impact on worker engagement, organizational learning and the achievement of safety goals. Unlike in the macro-case the emphasis is on attenuation. Figure 9.2 also shows how all this is done against the backdrop of organizational culture.

Team Psychological Safety

The dominant theme of team psychological safety in the literature is biased toward the examination of the construct in relation to subordinate-superior relationships. This research shows the plausibility of including in the literature an emphasis on similar dynamics between leaders at the same level – where reduced team psychological safety affects lateral communication, a condition symptomatic of group think.

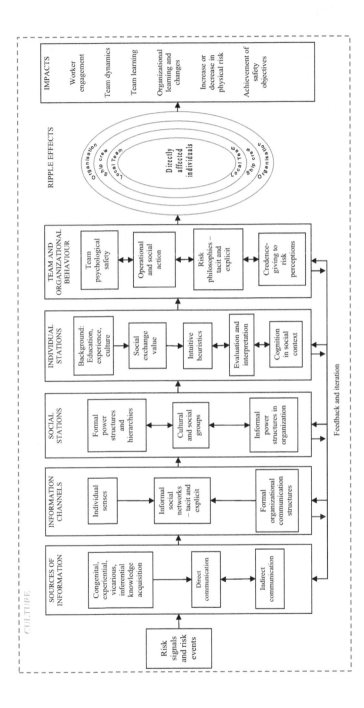

Figure 9.2 Micro-level adaptation of SARF

Methodological Implications and Recommendations

The research approach utilized here was one using mixed-methods but not including action research. The results show how a quantitative study was used to give underlying importance of variables and constructs and how a follow-up qualitative study could examine how the constructs were dynamically treated in shipping companies. It is believed that the results confirm the suitability of mixed-method research in certain areas in the maritime industry and shows how not constraining oneself to a particular research paradigm can help give a good overview of the area of inquiry. Though there are grounds in generalizing the findings of this research, any attempts to do so must be done with the following *proviso*s in mind:

- The interpretation of the data relating to nationality must not be seen as being suggestive of uniformity in national culture in a functionalist worldview. In keeping with the theory of structuration it is recognized that there is considerable social and individual forces at play in determining behaviours and interactions in any social context and at any given time.
- To avoid ecological fallacy, it must be appreciated that the data results – especially from the quantitative analyses – are statistical averages which may serve as guides to management in the prediction of behaviour but not as the basis for stereotypical determination or judgements of individual behaviour. There is enough research support for the existence of intra-national cultural diversity and individual personalities to indicate the inappropriateness of excessive stereotyping.

Despite the merits of the research approach used, it is noteworthy that there were some calls from research participants which seem to expect, even desire, a research approach that is best covered by action research.

43-year-old 2nd Officer – Male

We are also hoping that this survey will help the improvement in regards to the relations of the company and the seafarer. Thank you very much for doing this because by this survey our intention will be heard by our company. Thank you Sir!

55-year-old 2nd Engineer – Male

I am thanksfull [sic] to the questionnaire you give us for more development and upgrading us at sea and looking forward your assistance for us during our work at sea.

24-year-old 3rd Engineer – Male

Is the questionnaire going to have a positive influence on the way companies do their business?

36-year-old Chief Officer – Male

[There should be pressure] on charterers to insist on the highest possible standard from their 'charterees' since that is the most likely way to press home service quality delivery.

Action research is a relatively new research approach that pursues understanding and change *at the same time* often in an iterative and reflexive manner (Cunningham, 1993; Zuber-Skerritt, 1996).[4] Thus issues of axiology, which is not extremely pronounced in most methodologies, become prominent in action research because of a pronounced normative research approach. The mixed-methodology approach undertaken here had no iterative/reflexive change elements as far as the research itself goes and concludes with recommendations that are calls for improvement.

An action research approach would have endeavoured to make the change part of the research. It is recommended that further research consider the merits of this.

The need for a cross-disciplinary approach to studies in organizational learning and risk construal is recognized. The approach in this work has therefore been eclectic drawing from different theoretical domains. This author agrees with Berthoin Antal, Child, Dierkes and Nonaka (2001a, p. 936) when they recommend that:

More studies be conducted in mixed teams and that the research be designed in ways that enable multiple views of reality to be treated. In effect, researchers themselves need to learn how to learn better – in other words, they need to apply some of the lessons from the study of organizational learning to their own research processes.

Further Research

The research done in this work can be augmented by a replication using a larger sample size with each strata of the sample having a significant size. When analysing the data for nationality for example, one could only be restricted to differences between OECD and non-OECD because of the sample sizes involved. It would have been possible to explore more specific differences had the sample sizes for the different demographics (nationality/region) been greater. Further results that

---

4   See also http://users.monash.edu.au/~gromeo/action/lecnotes.rtf retrieved 24 June 2009.

increase these sample sizes will make it possible to generalize findings to a greater extent, with respect to nationalities, than has been possible in this work.

It is recommended that future research seek to synthesize from the survey items a concise list of items – based on factor analysis or principal component analysis – that can be used as a quick assessment tool in organizations to measure the constructs covered in the questionnaire. This kind of analysis should address the high level of shared variance between the constructs of team psychological safety and worker engagement on one hand and leader inclusiveness on the other.

Another potentially significant research direction would be applying the questionnaire items to ship officers in a single company and then investigating in-depth in that single company with interviews and so on. If this is repeated across many companies, there will be significant data for critiquing/augmenting the findings of this research and for exploring patterns of data as indicated by companies with high levels of organizational learning compared to those with low levels of organizational learning.

Appendix 1: Glossary

Accident

A sudden, unintended event or series of events where significant harm is inflicted on humans, the environment or material assets ... thus excludes intentional harm, such as terror, hacking or sabotage as well as harm that occurs gradually, such as long-term effects of continual emissions of toxic substances (Rosness et al., 2004).

Accidents involve damage to subsystems or the system as a whole, stopping the intended output or affecting it to the extent that it must be halted promptly (Perrow, 1999, p. 70).

An unwanted transfer of energy, because of lack of barriers and/or controls, producing injury to persons, property, or process, preceded by sequences of planning and operational errors, which failed to adjust to changes in physical or human factors and produced unsafe conditions and/or unsafe acts, arising out of the risk in an activity, and interrupting or degrading the activity (W.G. Johnson, 1980, p. 507).

Amplenuation

A neologism combining the words *amplification* and *attenuation*. It refers to the increasing or decreasing of risk signals based on filters of communication and perception such as social standing, ranking, education, context, control parameters, value and time.

Culture

A dynamic, intangible and composite system of interacting values, basic assumptions and norms which manifests in and influences individual and group attitudes, beliefs, behavioural patterns and non-behavioural items and which informs the meaning individuals/groups attribute to such manifestations in themselves and others.

The collective programming of the mind that distinguishes the members of one group or category of people from others (Hofstede, 2006).

Patterned ways of thinking, feeling and reacting, acquired and transmitted mainly by symbols, constituting the distinctive achievements of human groups, including their embodiments in artifacts; the essential core of culture consists of traditional (that is, historically derived and selected) ideas and especially their attached values (Kluckhohn, 1951, p. 86 as cited by Hofstede, 2001).

A pattern of shared basic assumptions that was learned by a group as it solved its problems of external adaptation and internal integration, that has worked well enough to be considered valid and, therefore, to be taught to new members as the correct way you perceive, think and feel in relation to those problems (Schein, 2004, p. 17).

Drift (also 'migration towards the boundary')

The natural migration of organizational activities towards the boundary of acceptable performance as a result of human behaviour in any work system being shaped by diverse objectives and constraints (Rasmussen, 1997, pp. 189–90).

Entropy

The degradation of an organization's system factors: processes, technology, the physical environment and human resources (Mol, 2003, p. xxi).

Leader Inclusiveness (LI)

Words and deeds by a leader or leaders that indicate an invitation and appreciation for others' contributions. [Leader inclusiveness] captures attempts by leaders to include others in discussions and decisions in which their voices and perspectives might otherwise be absent (Nembhard and Edmondson, 2006, p. 947).

Learning Organization

An organization skilled at creating, acquiring, and transferring knowledge, and at modifying its behaviour to reflect new knowledge and insight (Garvin, 1993).

An organization that encourages and facilitates the learning and development of people [individually and collectively and with respect to the strategic goals of the organization in the context of the external environment] at all levels of the organization, values [and retains] the learning and simultaneously transforms itself (Mullins, 2005, p. 1057).

An organization in which 'people at all levels, individually and collectively, are continually increasing their capacity to produce results they really care about' (Ratner, 1997 as cited in Cors, 2003, p. 9).

An organization that has an enhanced capacity to learn, adapt, and change. It's an organization in which learning processes are analyzed, monitored, developed, managed, and aligned with improvement and innovation goals. Its vision, strategy, leaders, values, structures, systems, processes, and practices all work to foster people's learning and development and to accelerate systems-level learning (Gephart, Marsick, Van Buren and Spio, 1996, pp. 36, 38).

A learning organization is one that learns continuously and transforms itself. Learning is a continuous, strategically used process – integrated with and running parallel to work (Marsick and Watkins, 2003, p. 142).

Organizations where people continually expand their capacity to create the results they truly desire, where new and expansive patterns of thinking are nurtured, where collective aspiration is set free, and where people are continually learning how to learn together (Senge, 2006, p. 3).

Near-miss

Any event that could have had bad consequences, but did not (Reason, 1997, p. 118).

Organizational Learning

The intentional or unintentional processes by which organizational social systems (made up of individuals and groups) create the potential (in cognitive, behavioural and/or affective terms) to adapt in behaviour, values and attitudes as a result of exposure to experiential, vicarious or contextual events whether or not there is a conscious awareness of this process. Learning also includes the process of hypothetically/rationally inferring causal links not explicitly based on historicity but linked to patterns stored in memory.

The process by which the organization constantly questions existing product, process and system, identify strategic position, apply various modes of learning, to achieve sustained competitive advantage (Wang and Ahmed, 2002).

The growing insights and successful restructurings of organizational problems by individuals reflected in the structural elements and outcomes of the organization itself (Simon, 1969 as cited in Fiol and Lyles, 1985, p. 803).

The activity and the process by which organizations eventually reach the ideal of a learning organization (Finger and Brand, 1999, p. 136).

The process of improving actions through better knowledge and understanding (Fiol and Lyles, 1985, p. 803).

The acquisition of new knowledge by actors who are able and willing to apply that knowledge in making decisions or influencing others in the organization (Miller, 1996, p. 486).

[Writing about organisms and organizations alike] ... An entity learns if, through its processing of information, the range of its potential behaviors is changed (Huber, 1991, p. 89).

Organizational learning occurs through shared insights, knowledge, and mental models ... Learning builds on past knowledge and experience – that is, on memory (Stata, 1989, p. 64).

Organizations are seen as learning by encoding inferences from history into routines that guide behaviour (Levitt and March, 1988, p. 319).

Organizational learning occurs when individuals within an organization experience a problematic situation and inquire into it on the organization's behalf. They experience a surprising mismatch between expected and actual results of action and respond to that mismatch through a process of thought and further action that leads them to modify their images of organization or their understanding of organizational phenomena and to restructure their activities so as to bring outcomes and expectations into line, thereby changing organizational theory-in-use. In order to become organizational, the learning that results from organizational inquiry must become embedded in the images of organization held in its members' minds and/or in the epistemological artefacts (the maps, memories, and programs) embedded in the organizational environment (Argyris and Schön, 1996, p. 16).

The way firms build, supplement and organize knowledge and routines around their activities and within their cultures, and adapt and develop organizational efficiency by improving the use of broad skills in their workforces (Dodgson, 1993, p. 377).

Precursor

That which runs or goes before ... precedes and heralds the approach of another (Murray et al., 1991).

A generic, undesirable but stable condition that is an antecedent of accidents and is not dependent on chance in the sense in which near-misses are. A precursor has the potential of contributing to the kinds of dynamics that lead to near-misses and accidents and thus heralds both accidents AND near-misses. It does not rely for its differentiation from accidents on outcomes.

Recreancy

A word 'based on the Latin roots *re-* (back) and *credere* (to entrust). This usage draws on one of the two dictionary meanings of the term, namely a retrogression or failure to follow through on a duty or trust. It is intended to provide an affectively neutral reference to behaviors of persons and/or of institutions that hold positions of trust, agency, responsibility, or fiduciary or other forms of broadly expected obligations to the collectivity, but that behave in a manner that fails to fulfill the obligations or merit the trust' (Freudenburg, 1993, pp. 916–17).

Resilience (with respect to an organization)

The capacity of an organization to accommodate failures and disturbances without producing serious accidents (Rosness et al., 2004, p. 9).

Team Psychological Safety (TPS)

'The shared belief held by members of a team that the team is safe for interpersonal risk-taking' (Edmondson, 1999, p. 350) and where these members feel 'able to show and employ [themselves] without fear of negative consequences to self-image, status, or career' (Kahn, 1990, p. 708).

Worker Engagement

The harnessing of organization members' selves to their work roles; in engagement, people employ or express themselves physically, cognitively, and emotionally during role performance (Kahn, 1990, p. 694).

Individuals' connection to work intellectually, emotionally, creatively and psychologically (Krause, 2000, p. 24).

The development of relationships between workers and employers based on collaboration and trust and nurtured as part of the management of health and safety. This goes further than simply consulting workers. It involves a commitment to solving problems together (Health and Safety Commission, 2006, p. 4).

Appendix 2:
Final Administered Survey

Organizational Learning in Shipping

Survey

Part of a study into the Social Amplification of Risk in Shipping Organizations and associated effects on Organizational Learning
    After reading the accompanying introduction, PLEASE START HERE:

- Are you currently an officer of a seagoing ship or have you been at sea as an officer in the last five years?

    ☐ Yes (please continue to Section A)

    ☐ No
    – (If No) It is not necessary for you to complete the remainder of this questionnaire. Thank you for your willingness to help.

Section A:

In this section we would like you to tell us about the team you work with on board (or with ship–shore management) – for example, bridge team, engine room team or company operations management. Please indicate your agreement/ disagreement with the following statements, by ticking the appropriate box. *If you are not presently on board, kindly think of your last team on your last ship.*

	Strongly agree	Agree	Neutral	Disagree	Strongly disagree
A1 Seafarers in this team are able to talk openly about difficult operational problems with each other.	☐	☐	☐	☐	☐

A2 I feel that pointing out what I consider risky regarding work is acceptable to others in the team.	☐	☐	☐	☐	☐
A3 I sometimes hesitate in contributing to the work of the team because of how people in the team shift blame on others.	☐	☐	☐	☐	☐
A4 Working in this team, my unique skills and talents are valued and utilized.	☐	☐	☐	☐	☐
A5 During routine shipboard work, I am able to freely share my opinions regarding safety with my work colleagues.	☐	☐	☐	☐	☐
A6 If I encounter a problem at work I will not hesitate to ask my superior for help.	☐	☐	☐	☐	☐

A7 In my company and regarding safety at work, if I disagree with a decision or intended course of action by a superior I will …

	Strongly agree	**Agree**	**Neutral**	**Disagree**	**Strongly disagree**
… remain silent.	☐	☐	☐	☐	☐
… state my doubts to another team member.	☐	☐	☐	☐	☐
… state my doubts to the superior officer.	☐	☐	☐	☐	☐
Other (please state).					

A8 You are welcome to put down any other comments you may have about how you feel as a member of this team especially regarding speaking up concerning safety and maritime pollution.

_____

_____

_____

_____

_____

Section B:

In this section we would like you to tell us about the leader of your team. *If you are not on board now, kindly think about the leader of your last team on your last ship.* The team refers to the group you work with on board (for example, on the bridge or in the engine room). For Masters and Chief Engineers please consider your team as the operational management team on board and ashore.

- Whom do you consider as your team leader?

  ☐ Master

  ☐ Chief Engineer

  ☐ Other (please specify): _____

Kindly indicate your agreement/disagreement with the following statements:

	Strongly agree	Agree	Neutral	Disagree	Strongly disagree
B1 The team leader initiates meetings to discuss the team's work progress.	☐	☐	☐	☐	☐
B2 It is difficult to approach the team leader about work related problems.	☐	☐	☐	☐	☐
B3 The team leader acknowledges it when (s)he is wrong.	☐	☐	☐	☐	☐
B4 When someone makes a mistake in this team, the leader helps the team to learn from it.	☐	☐	☐	☐	☐

B5 The team leader does NOT value the opinion of others.	☐	☐	☐	☐	☐
B6 The team leader actively seeks (encourages) feedback.	☐	☐	☐	☐	☐
B7 The team leader acknowledges feedback.	☐	☐	☐	☐	☐
B8 The team leader sometimes asks for help.	☐	☐	☐	☐	☐
B9 The team leader encourages asking for help.	☐	☐	☐	☐	☐

B10   You are welcome to put down any other comments you may have about your team leader's leadership of your team.

_____

_____

_____

_____

_____

Section C:

Thinking about the company you work for (*or last worked for if you are not on board now*) kindly indicate your agreement or disagreement with the following statements:

	Strongly agree	Agree	Neutral	Disagree	Strongly disagree
C1 This company encourages learning from accidents.	☐	☐	☐	☐	☐
C2 This company does **not** welcome the questioning of the way things have always been done.	☐	☐	☐	☐	☐

C3 This company often develops new ways of doing safety-related things, based on past experience.	☐	☐	☐	☐	☐
C4 This company spends time making sure that every team member understands the company's objectives.	☐	☐	☐	☐	☐
C5 Communication channels in this company are effective.	☐	☐	☐	☐	☐
C6 When things about safety go wrong, reasons for this are examined in a constructive manner rather than seeking to blame someone.	☐	☐	☐	☐	☐
C7 In this company, there is a culture of continuous improvement where people are always trying to learn and work better.	☐	☐	☐	☐	☐
C8 In this company ways of working and learning are recorded and stored in an accessible way..	☐	☐	☐	☐	☐
C9 This company encourages the team on board to take the initiative in making improvements in our work areas.	☐	☐	☐	☐	☐
C10 In this company, we are encouraged to learn while we are working.	☐	☐	☐	☐	☐
C11 When a safety procedure is replaced by another, I know why.	☐	☐	☐	☐	☐

C12 You are welcome to make any other comments you may have about how the company learns from its seafarers.

_____

_____

_____

_____

Section D:

Considering the shipping company you work with, your shipboard team and you as an individual member (*now or on your last ship*), please indicate your agreement/disagreement with the following statements:

	Strongly agree	Agree	Neutral	Disagree	Strongly disagree
D1 There are procedures by which this team can contribute (make suggestions) to the development of safety in this company.	☐	☐	☐	☐	☐
D2 Shore management actively seeks this team's suggestions on matters regarding safety.	☐	☐	☐	☐	☐
D3 My team members are committed to doing the best job they can.	☐	☐	☐	☐	☐
D4 I know what is expected of me at work.	☐	☐	☐	☐	☐
D5 I trust this company in its ability to consider my perceptions of risks on board ship.	☐	☐	☐	☐	☐
D6 Everything I need to do my work right is available to me.	☐	☐	☐	☐	☐
D7 I often get recognition for a good job done.	☐	☐	☐	☐	☐
D8 I feel that my work on board is appreciated by the company as being important.	☐	☐	☐	☐	☐
D9 The company gives me opportunities for learning and advancing my career.	☐	☐	☐	☐	☐

D10 Company management gives me the opportunity to give my views on risk policy (input to its development or changes).	☐	☐	☐	☐	☐
D11 Company management gives me the opportunity to give my views on safety problems that exist on board.	☐	☐	☐	☐	☐

D12 What influence do you think you have over shore management's decisions regarding on board risk? (Please choose the most applicable one)

☐ You have no influence on management decisions.

☐ Your opinion is rarely considered by management.

☐ Your opinion is sometimes considered by management.

☐ Your opinion is always considered by management.

☐ You have power to change a management decision.

☐ Not sure.

D13 Suppose your company is provided with information about how a given seafarer views some specific risk. When the importance of this information is assessed by management, how much weight do you think the company will assign to each of the following characteristics of that person: (please rank by a number from 0 to 10, with 10 being highest importance)

Rank of the person.	
Age of the person.	
Gender of the person.	
Nationality of the person.	
How long the person has been in the company.	

The person's informal relations with others in the company.	
The person's experience from other companies.	
Other (please specify).	

Section E:

E1   What is your age? _____ years

E2   What is your nationality? _____

E3   What is your gender? ☐ Male   ☐ Female

E4   What is your rank on board (or your last rank if you are off the ship now)?_

E5   For how long have you worked with your current company?
_____ years

E6   Do you find your present income satisfactory?

       ☐ Yes

       ☐ No

E7   How were you employed?

       ☐ Through a manning agency

       ☐ Directly by the shipping company

       ☐ Other (please specify) _____

E8   What is the nature of your employment contract with the company?

       ☐ Short-term contract (renewable)

       ☐ Short-term contract (not to be renewed)

       ☐ Long-term contract (permanent)

       ☐ Other: (please describe) _____

E9   How many years have you worked …

At sea? _____

In shore side ship management? _____

E10   Have you ever been employed as a rating?

☐ Yes

☐ No

E11   During your time of working at sea have you worked on board ship with a multinational crew?

☐ Yes

☐ No

E12   How big is the fleet of the company in which you work (or last worked)?

☐ Below 5 vessels

☐ 5–20 vessels

☐ Above 20 vessels

E13   In which country have you had most of your work-related training (school)? _____

E14   What was the flag of your last ship (port of registry)? _____

E15   Do you enjoy:

Working as a seafarer? ☐ Yes ☐ No

Please state why: _____
_____

Working with your present company? ☐ Yes ☐ No

Please state why: _____
_____

E16   Please indicate in the table below:

   a. the ship type you are working on (or the most recent if you are not on board now), and

   b. the ship type on which you have spent most time at sea.

		Please tick ONE in each column	
		Present (or *LAST* ship)	**Type on which I have spent most time at sea**
1	Gas Tanker	☐	☐
2	Chemical Tanker	☐	☐
3	Oil Tanker	☐	☐
4	Other Tanker	☐	☐
5	OBO Oil/Bulk Dry	☐	☐
6	Bulk Carrier	☐	☐
7	General Cargo	☐	☐
8	Container Vessel	☐	☐
9	Reefer	☐	☐
10	Ro-Ro Cargo/Car Carrier	☐	☐
11	Passenger Ro-Ro	☐	☐
12	Passenger Cruise Ship	☐	☐
13	Livestock	☐	☐
14	Offshore supply	☐	☐
15	Research vessel	☐	☐
16	Dredger	☐	☐
17	Other (please specify) _____	☐	

Thank you very much for taking the time to complete this questionnaire. Your help in providing this information is very much appreciated. If there is anything else you would like to tell us to help us understand further how shipping companies learn from their seafarers, kindly do so in the space below. You may also put down any comments you have about this questionnaire (*for example questions you may have been expecting but did not see*).

_____
_____
_____
_____
_____
_____
_____
_____
_____
_____

Appendix 3:
Demographic Variables for Survey Data Set

Table A.1   **Demographics categorical variables**

	Value	Count	Per cent
	OECD	28	23.0
	Eastern/Central Europe	19	15.6
	Africa/Middle East	9	7.4
Nationality	Latin America	0	0.0
	Caribbean	0	0.0
	Far East	52	42.6
	Indian sub-continent	13	10.7
	Missing values	1	0.8
Gender	Female	4	3.3
	Male	118	96.7
Rank	Master	26	21.3
	Chief Officer	19	15.6
	Operational Level Deck Officer	46	37.7
	Chief Engineer	9	7.4
	2nd Engineer	8	6.6
	Operational Level Eng. Officer	10	8.2
	Other	2	1.6
	Missing Values	2	1.6
	OECD	28	23.0
	Eastern/Central Europe	20	16.4
	Africa/Middle East	8	6.6
Respondent's country of training	Latin America	1	0.8
	Caribbean	0	0.0
	Far East	41	33.6
	Indian sub-continent	13	10.7
	Multi/various	4	3.3

Table A.1   *Continued*

Satisfied with income	No	40	32.8
	Yes	82	67.2
Experience as rating	No	43	35.2
	Yes	78	63.9
	missing	1	0.8
Experience with multinational crew	No	9	7.4
	Yes	113	92.6
Mode of employment	Manning agency	60	49.2
	Directly by shipping company	58	47.5
	Other	4	3.3
Nature of employment contract	Short-term contract renewable	69	56.6
	Short-term contract not renewable	12	9.8
	Long-term contract permanent	40	32.8
	Missing values	1	0.8
Employing company fleet size	Below 5 vessels	12	9.8
	5–20 vessels	36	29.5
	Above 20 vessels	74	60.7
Enjoy working at sea	No	24	19.7
	Yes	96	78.7
	Missing	2	1.6
Enjoy working in present company	No	19	15.6
	Yes	98	80.3
	Missing	5	4.1

Table A.1   *Continued*

Team leader	Master	89	73.0
	Chief Engineer	21	17.2
	Other	5	4.1
	Missing	7	5.7
Fleet size	Below 5 vessels	12	9.8
	5–20 vessels	36	29.5
	Above 20 vessels	74	60.7
Rank grouping	Operational level	58	47.5
	Management level	62	50.8
	Missing values	2	1.6
Age grouping	<= 30	42	34.4
	31–43	41	33.6
	44+	39	32.0
On board department	Deck department	91	74.6
	Engine department	27	22.1
	Other	2	1.6
	Missing values	2	1.6

Table A.2   Demographics continuous variables

	Valid	Missing	Mean	SD	Min	Max
Age	122	0	37.02	11.34	21	62
Time in present company	119	3	4.68	5.95	0.2	35.0
Time at sea	121	1	13.21	10.34	0.5	42.0
Time in shore management	122	0	0.49	1.92	0.0	16.0

Table A.3   Flag of respondent's (last) ship

	Count	Per cent
Missing	2	1.6
Antigua	1	0.8
Bahamas	8	6.6
Cambodia	2	1.6
Cyprus	1	0.8
Faroe Islands	1	0.8
Germany	1	0.8
Greece	3	2.5
Hong Kong	7	5.7
Japan	1	0.8
Jordan	1	0.8
Liberia	10	8.2
Malta	16	13.1
Marshall Islands	2	1.6
Nigeria	2	1.6
Norway	2	1.6
Panama	24	19.7
Portugal	2	1.6
Singapore	3	2.5
South Korea	11	9.0
St Kitts and Nevis	2	1.6
Sweden	19	15.6
UK	1	0.8

Table A.4   Respondent's (last) ship

	Count	Per cent
Missing	5	4.1
Gas tanker	9	7.4
Chemical tanker	8	6.6
Oil tanker	13	10.7
Bulk carrier	28	23.0
General cargo	11	9.0
Container vessel	12	9.8
Reefer	1	0.8
Roro cargo / car carrier	23	18.9
Passenger cruise ship	1	0.8
Offshore supply	4	3.3
Research vessel	4	3.3
Other	3	2.5

Appendix: 4
Survey Scales and Inter-Item Correlation Matrices

Table A.5   Team psychological safety: Inter-item correlation matrix

		2	3	4	5	6	7	8	9
1	Able to talk openly	.537	.226	.343	.552	.284	.327	.215	.304
2	Able to point out risk	1.000	.279	.272	.409	.226	.345	.142	.192
3	Presence of blame game		1.000	.174	.149	.260	.340	-.047	.210
4	My unique skills valued			1.000	.551	.441	.372	.057	.308
5	Freely share safety opinions				1.000	.316	.398	.167	.315
6	Able to ask superior for help					1.000	.358	-.009	.300
7	Silence						1.000	-.061	.471
8	Doubts to team member							1.000	.223
9	Doubts to superior								1.000

Table A.6   Leader inclusiveness

		2	3	4	5	6	7	8	9
1	Initiates meetings	.065	.426	.568	.415	.622	.575	.237	.200
2	Difficult to approach	1.000	.087	.195	.478	.250	.320	.244	.107
3	Acknowledges wrong		1.000	.381	.312	.437	.449	.284	.385
4	Use mistakes for team learning			1.000	.394	.536	.519	.279	.208
5	Does NOT value others' opinions				1.000	.475	.497	.267	.266
6	Encourages feedback					1.000	.650	.362	.316
7	Acknowledges feedback						1.000	.248	.205
8	Asks for help							1.000	.205
9	Encourages asking for help								1.000

Table A.7   Organizational learning

		2	3	4	5	6	7	8	9	10	11
1	Encourages learning from accidents	.241	.470	.425	.427	.399	.563	.406	.499	.513	.275
2	Does NOT welcome questioning	1.000	.350	.393	.317	.317	.411	.171	.414	.385	.417
3	Develops new ways		1.000	.572	.635	.439	.730	.439	.602	.605	.275
4	Company's objectives understood			1.000	.679	.595	.664	.520	.684	.652	.260
5	Communication channels effective				1.000	.534	.720	.553	.545	.544	.187
6	No blame culture					1.000	.573	.378	.525	.527	.209
7	Continuous improvement						1.000	.584	.711	.634	.254
8	Ways recorded and stored							1.000	.572	.469	.103
9	Encourages initiatives								1.000	.698	.230
10	Encourages learning with work									1.000	.279
11	Know why old replaced										1.000

Table A.8   Worker engagement

		2	3	4	5	6	7	8	9	10	11	12
1	Procedures for contributing	.399	.324	.208	.321	.340	.297	.414	.329	.394	.324	.184
2	Management actively seeks suggestions	1.000	.419	.177	.456	.348	.406	.498	.476	.550	.588	.252
3	Team members committed		1.000	.458	.426	.378	.346	.436	.411	.366	.422	.187
4	Know what is expected of me			1.000	.363	.237	.165	.215	.206	.152	.279	.236
5	Trust company				1.000	.459	.439	.568	.476	.434	.517	.362
6	Everything I need available					1.000	.445	.398	.389	.382	.319	.293
7	Recognition for good job						1.000	.594	.507	.405	.365	.292
8	Work appreciated by company							1.000	.599	.555	.618	.413
9	Opportunities for advancement								1.000	.657	.599	.236
10	Opportunity for views on risk policy									1.000	.669	.385
11	Opportunity for views on safety o/b										1.000	.373
12	Influence over decisions											1.000

Appendix 5:
Statistical Data for Computed
Construct Variables

Table A.9   Tests of normality for computed variables

	Kolmogorov-Smirnov†			Shapiro-Wilk		
	Statistic	df	Sig.	Statistic	df	Sig.
TPS	.057	122	.200††	.986	122	.253
LI	.104	120	.003	.974	120	.022
OL	.107	121	.002	.970	121	.008
WE	.114	121	.001	.975	121	.023

† Lilliefors significance correction.

†† This is a lower bound of the true significance.

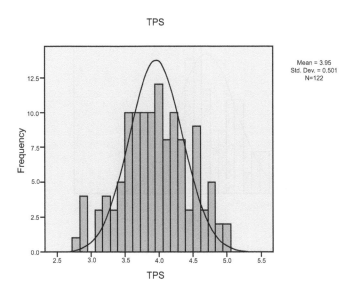

Figure A.1   Frequency distribution of TPS results

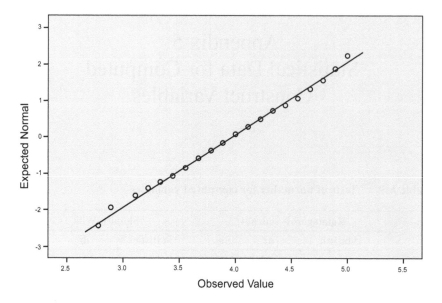

Figure A.2   Normal Q–Q plot of TPS

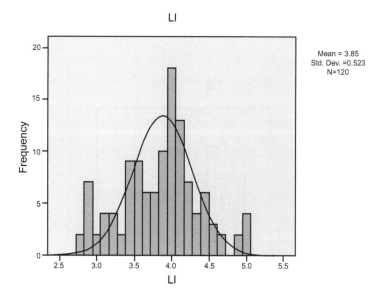

Figure A.3   Frequency distribution of LI results

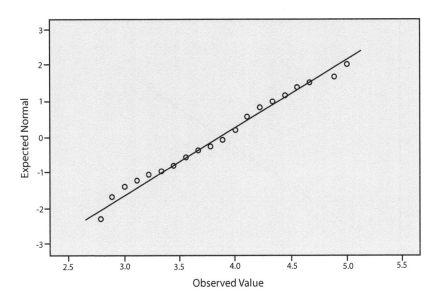

Figure A.4   Normal Q–Q plot of LI

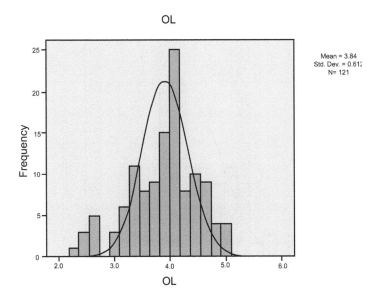

Figure A.5   Frequency distribution of OL results

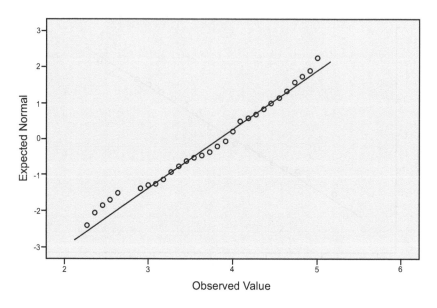

Figure A.6   Normal Q–Q plot of OL

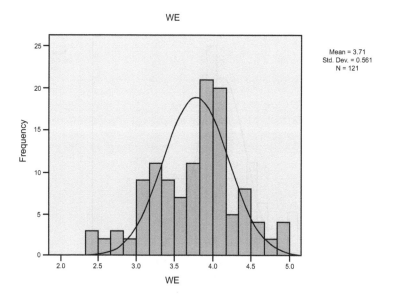

Figure A.7   Frequency distribution of WE results

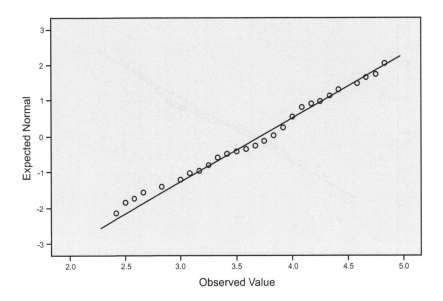

Figure A.8   Normal Q–Q plot of WE

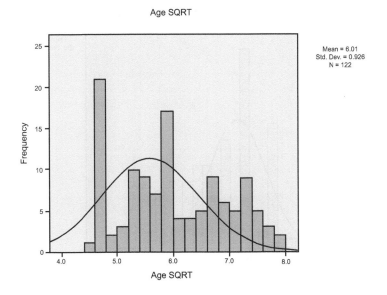

Figure A.9   Frequency distribution of square root transformation of age

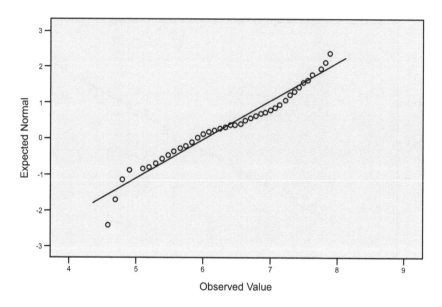

Figure A.10   Normal Q–Q plot of square root transformation of age

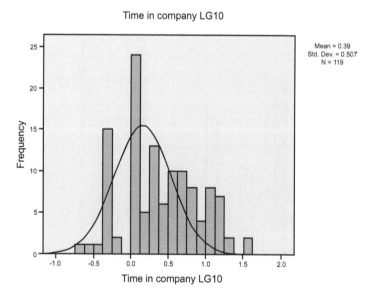

Figure A.11   Frequency distribution of log transformation of time in company

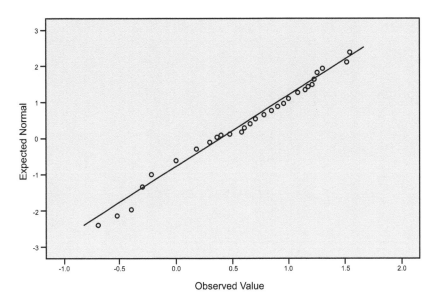

Figure A.12 Normal Q–Q plot of log transformation of time in company

Appendix 6:
Results of Statistical Tests

Table A.10 **Model summary of standard multiple regression analysis of TPS, LI and WE as predictors of OL (dependent variable)**

Model	R	R square	Adjusted R square	Std. error of the estimate
1	.781[†]	.609	.599	.3873

[†] Predictors: (Constant), WE, TPS, LI.

Table A.11 **ANOVA of standard multiple regression analysis of TPS, LI and WE as predictors of OL (dependent variable)**

Model		Sum of squares	df	Mean square	F	Sig.
1	Regression	26.903	3	8.968	59.785	.000[†]
	Residual	17.250	115	.150		
	Total	44.152	118			

[†] Predictors: (Constant), WE, TPS, LI.

Table A.12 Coefficients and correlations of standard multiple regression analysis of TPS, LI and WE as predictors of OL (dependent variable)

	Unstandardized coefficients		Standardized coefficients			95.0% Confidence interval for B		Correlations			Collinearity statistics	
	B	Std. error	Beta	t	Sig.	Lower bound	Upper bound	Zero-order	Partial	Part	Tolerance	VIF
(Constant)	.086	.309	+	.277	.782	-.527	.698					
TPS	.385	.092	.315	4.170	.000	.202	.567	.626	.362	.243	.596	1.678
LI	-.025	.100	-.021	-.246	.806	-.222	.173	.559	-.023	-.014	.468	2.139
WE	.626	.090	.574	6.947	.000	.448	.805	.738	.544	.405	.497	2.012

Table A.13 Coefficients and correlations of hierarchical multiple regression[†] analysis of TPS and WE as predictors of OL, controlling for rank, age and time in company

Model	Unstandardized coefficients B	Std. Error	Standardized coefficients Beta	t	Sig.	95.0% Confidence interval for B Lower bound	Upper bound	Correlations Zero-order	Partial	Part	Collinearity statistics Tolerance	VIF
1 (Constant)	2.173	.370		5.872	.000	1.440	2.906					
Rank grouping	.121	.118	.099	1.019	.310	-.114	.355	.331	.095	.083	.704	1.420
Age (SQRT)	.261	.067	.396	3.899	.000	.129	.394	.479	.343	.318	.645	1.550
Time in company (LG10)	.082	.113	.068	.728	.468	-.142	.306	.289	.068	.059	.759	1.318
2 (Constant)	-.498	.365		-1.364	.175	-1.222	.226					
Rank grouping	-.067	.084	-.055	-.800	.426	-.234	.100	.331	-.075	-.045	.668	1.497
Age (SQRT)	.152	.047	.230	3.201	.002	.058	.246	.479	.290	.181	.618	1.619
Time in company LG10	-.055	.079	-.045	-.689	.493	-.212	.103	.289	-.065	-.039	.739	1.353
TPS	.337	.088	.276	3.835	.000	.163	.511	.626	.341	.217	.617	1.621
WE	.578	.076	.530	7.568	.000	.427	.729	.738	.582	.427	.650	1.538

[†] Dependent variable: OL.

Table A.14   Model summary[†] of hierarchical multiple regression analysis of TPS and WE as predictors of OL, controlling for rank, age and time in company

Model	R	R square	Adjusted R square	Std. error of the estimate	Change statistics R square change	F change	df1	df2	Sig. F change
1	.492[††]	.242	.222	.5395	.242	12.138	3	114	.000
2	.802[†††]	.643	.627	.3737	.401	62.785	2	112	.000

[†] Dependent Variable: OL.

[††] Predictors: (Constant), Time in company LG10, Rank grouping, Age SQRT.

[†††] Predictors: (Constant), Time in company LG10, Rank grouping, Age SQRT, WE, TPS.

Table A.15   ANOVA[†] of hierarchical multiple regression analysis of TPS and WE as predictors of OL, controlling for rank, age and time in company

Model		Sum of squares	df	Mean square	F	Sig.
1	Regression	10.598	3	3.533	12.138	.000[††]
	Residual	33.180	114	.291		
	Total	43.778	117			
2	Regression	28.136	5	5.627	40.291	.000[†††]
	Residual	15.642	112	.140		
	Total	43.778	117			

[†] Dependent Variable: OL.

[††] Predictors: (Constant), Time in company LG10, Rank grouping, Age SQRT.

[†††] Predictors: (Constant), Time in company LG10, Rank grouping, Age SQRT, WE, TPS.

Table A.16   Mann-Whitney U test: TPS scores for nationality, rank, multinational crew experience and rating experience

Test Variable: Team Psychological Safety

IV	Groups	n	Median	Mann-Whitney U	z	Asymptotic Sig. (2-tailed)	Effect size[†]
Nationality	Non-OECD	93	3.889	820.0	- 2.970	.003	0.23
	OECD	28	4.222				
Rank	Operational	58	3.778	1012.5	- 4.135	.000	0.38
	Management	62	4.111				
Experience with multinational crew	No	9	3.778	327.5	- 1.777	.076	
	Yes	113	4.000				
Experience as rating	No	43	4.000	1569.5	- 0.584	.559	
	Yes	78	3.889				

[†] Only indicated for significant results.

Table A.17   Mann-Whitney U test: LI scores for nationality, rank, multinational crew experience and rating experience

Test Variable: Leader Inclusiveness

IV	Groups	n	Median	Mann-Whitney U	z	Asymptotic Sig. (2-tailed)	Effect size[†]
Nationality	Non-OECD	91	4.000	1172.5	- 0.638	.523	
	OECD	28	3.889				
Rank	Operational	56	3.667	986.0	- 4.056	.000	0.37
	Management	62	4.000				
Experience with multinational crew	No	9	3.667	398.0	- 1.015	.310	
	Yes	111	3.889				
Experience as rating	No	43	3.889	1331.5	- 1.679	.093	
	Yes	76	4.000				

[†] Only indicated for significant results.

Table A.18   Mann-Whitney U test: WE scores for nationality, rank, multinational crew experience and rating experience

Test Variable: Worker Engagement

IV	Groups	n	Median	Mann-Whitney U	z	Asymptotic Sig. (2-tailed)	Effect size[†]
Nationality	Non-OECD	92	3.833	1196	- 0.572	.567	
	OECD	28	3.875				
Rank	Operational	58	3.458	1118.5	- 3.467	.001	0.32
	Management	61	3.917				
Experience with multinational crew	No	9	3.500	429.5	- 0.738	.461	
	Yes	112	3.833				
Experience as rating	No	43	3.750	1441.5	- 1.174	.240	
	Yes	77	3.833				

[†] Only indicated for significant results

Table A.19   Mann-Whitney U test: OL scores for nationality, rank, multinational crew experience and rating experience

Test Variable: Organizational Learning

IV	Groups	n	Median	Mann-Whitney U	z	Asymptotic Sig. (2-tailed)	Effect size[†]
Nationality	Non-OECD	92	3.955	1268.5	- 0.121	.903	
	OECD	28	3.955				
Rank	Operational	57	3.727	1133.5	- 3.379	.001	0.31
	Management	62	4.000				
Experience with multinational crew	No	9	3.727	407.5	- 0. 956	.339	
	Yes	112	4.000				
Experience as rating	No	43	3.818	1390.50	- 1.454	.146	
	Yes	77	4.000				

[†] Only indicated for significant results

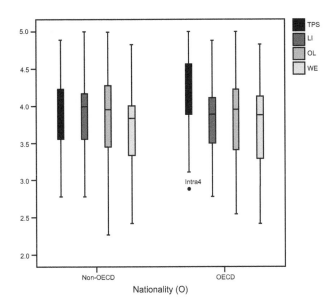

Figure A.13 Nationality median differences for constructs

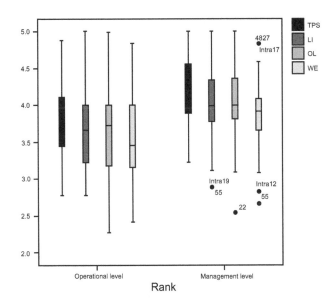

Figure A.14 Rank median differences for constructs

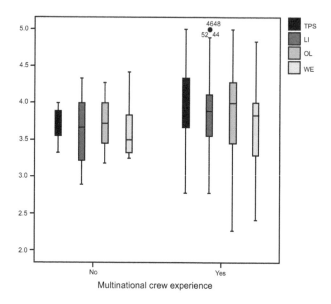

Figure A.15　Multinational experience median differences

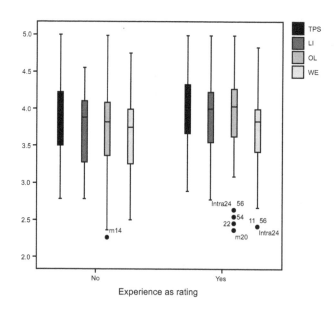

Figure A.16　Experience as rating median differences for constructs

Table A.20   Differences in age groups for TPS, LI, OL and WE values

Age Groups		TPS	LI	OL	WE
<= 30	N	42	40	41	42
	Median	3.667	3.556	3.455	3.333
31 – 43	N	41	41	41	41
	Median	4.111	4.000	4.000	4.000
44+	N	39	39	39	38
	Median	4.111	4.000	4.182	3.917
Total	N	122	120	121	121
	Median	4.000	3.889	3.909	3.833

Table A.21   Kruskal-Wallis test statistics[†]

	TPS	LI	OL	WE
Chi-Square	22.159	22.153	34.923	24.952
df	2	2	2	2
Asymp. Sig.	.000	.000	.000	.000

[†]Grouping Variable: Age (Binned)

Appendix 7:
Semi-Structured Interviews:[1]
Guidance Questions

1. Do you have a background in seafaring?
2. How long have you been working in this job? In what capacities over that period?
3. How are risks identified in your organization?
4. Does the organization classify risk in any way?
5. How are risks classified?
6. In your opinion is this the right way to classify risk?
7. Do you think risk should be classified, and if yes, how?
8. What are the sources of information/knowledge you have regarding risks to the operation of your ships?
9. What factors does your organization consider when assessing the authenticity (acceptability, credence-worthiness) of a risk perception (knowledge about risk)?
10. What are the organization's goals regarding quality?
11. What are the perceptions of management regarding the seafarers' attitudes, motivation and commitment to the company's goals regarding quality[2]?
12. What are your perceptions of the ability and willingness of your sea-staff to 'speak up'?
13. Do you have an institutional way of ascertaining/assessing this? If yes, how?
14. Have you ever encountered an incident of 'whistle-blowing' (internal or external)?
15. Do you think there is a culture of 'going along to get along' in this organization?
16. Do you have a formal voluntary reporting system such as CHIRP or SAFIR? What is its nature?
17. In your opinion what would constitute a nonconformity regarding quality?
18. Who is normally the originator of nonconformity reports?

---

1   The questions were used only as a guide. To a large extent, the interviewees were free to structure their own accounts, with few promptings to keep the accounts within the subject area.

2   Quality → Safe, secure and pollution-free operations of a very high standard.

19. Can you describe the process by which these nonconformity reports get to the attention of those able to rectify them?
20. How do you think Masters on your ships motivate their crew in the observance of the safety and environmental protection policy?
21. Does the company keep statistics of seafarer turnover? What is the retention rate over a period of say 5 years?
22. What factors do you think make a worker in your organization more engaged in organizational quality goals?
23. When (if) you consult the crew of a ship regarding risks, what is the scope of issues covered? Do they relate to purely physical hazards or do they extend to organizational management, safety culture, and so on?
24. In the hierarchy of risk controls, where does input from your seafarers rank? Avoid risk, reduce risk or PPE?
25. What in your opinion is management's understanding of the depth of accident causation?
26. What do you think is the seafarers' understanding of the depth of accident causation?
27. What systematic processes do you have that ensure that any feed-in from seafarers is recorded and tracked for attention and closure?
28. Do you have auditable processes that ensure that decisions made regarding risk have taken account of worker's input?
29. Have you ever had any in-house training regarding communication and teamwork in this organization?
30. Do you know of anybody (management and workers) who has had any such in-house training?
31. Can you give me two episodes of what you would consider as high-quality organizational learning?
32. Can you give me two episodes of what you would consider as low-level organizational learning?
33. How do different parts of your organization learn from the experiences of other parts?
34. Do you ever carry out training for teams together?
35. Do you try to get the same teams to work together often?
36. How, once it has been generated or accessed, is learning diffused throughout the organization and shared by organizational means?
37. How does shore management ensure the alignment of individual learning with organizational goals?
38. Have you encountered the notion of a 'learning organization'?
39. What does the construct of organizational learning mean to you?
40. How does organizational learning take place here?
41. In what way do you think your organization is a learning organization?
42. How would you say this organization determines the need to acquire new knowledge?
43. Can you tell me about the *Bow Mariner* accident?

44. Can you tell me of any accident that has happened in another shipping company in the last couple of years that has led to changes in this company?
45. Can you give me the number of claims this company has made regarding hull/machinery insurance and P&I over the last 5 years? Total value?

Appendix 8:
Qualitative Data Codes

Codes of Theoretical Themes

Knowledge Acquisition

1. Knowledge Acquisition: Congenital Learning
2. Knowledge Acquisition: Contextual Learning
3. Knowledge Acquisition: Experiential Learning
4. Knowledge Acquisition: Feedback Cycles
5. Knowledge Acquisition: Grafting
6. Knowledge Acquisition: Inferential Learning
7. Knowledge Acquisition: Searching and Noticing – Ship Visits
8. Knowledge Acquisition: Vicarious Learning/External Systems

Information Interpretation

9. Information Interpretation: Cognitive Maps and Framing
10. Information Interpretation: Data Analysis
11. Information Interpretation: Information Selection and Overload
12. Information Interpretation: Media Richness

Information Distribution

13. Information Distribution: Computer-based/paper-based
14. Information Distribution: Formal Meetings and Reports
15. Information Distribution: Professional Information Brokers

Organizational Memory

16. Organizational Memory: Computer-based/paper-based
17. Organizational Memory: Procedures and Routines
18. Organizational Memory: Retrieval of Information
19. Organizational Memory: Storage of Information
20. Organizational Memory: Transactive

Structural Change and Adaptation

Codes of Emergent Themes

1. Accident Causation Philosophy and Modelling
2. Action Research Call
3. Background of Interviewee
4. Behaviour Based Safety
5. Blame/Accountability
6. Communication Channels: Risk – Individual
7. Communication Channels: Welfare – Individual
8. Communication: Language
9. Communication: Ship – Shore
10. Competence and Training
11. Complacency
12. Continuous Improvement
13. Credence-giving to Risk Perceptions
14. Crew Engagement: Amenities on Board
15. Crew Engagement: Contracts and Crewing Links with Owners
16. Crew Engagement: Motivation
17. Crew Engagement: Professional Development
18. Crew Engagement: Salary
19. Crew Engagement: Social Relations Onboard
20. Crew Engagement: Social Security and Welfare
21. Culture: Nationality
22. Culture: Organizational
23. Culture: Professional
24. Entropy/Drift/Migration
25. Gender issues
26. Human Resource Retention
27. Individual/Collective
28. Inter-Organizational Information Sharing
29. Inter-Organizational Peer Pressure
30. Inter-Organizational Resource Sharing
31. Leader Inclusiveness
32. Learning Level: Deutero
33. Learning Level: Double-loop
34. Learning Level: Single-loop
35. Limitations to Learning: Legal
36. Limitations to Learning: Measurability
37. Limitations to Learning: Organizational Recreancy
38. Organizational Structure – Shipboard
39. Owner/Manager/Charterer Control
40. Precursor Analysis

41. Productivity/Safety Balance
42. Psychometric Testing
43. Resource Availability and Utilization
44. Risk Assessment
45. Risk Identification
46. Safety Administration Burden
47. Safety Attitudes
48. Safety Management System: Formal
49. Safety Management System: Informal
50. Speaking up
51. Team Psychological Safety
52. Teamwork

References

Abrahamson, E. (1996). Management fashion. *The Academy of Management Review, 21*(1), 254–85.

Adams, J.S. (1965). Inequity in social exchange. In L. Berkowitz (ed.), *Advances in experimental social psychology* (vol. 2, pp. 267–99). New York: Academic Press.

Air Accidents Investigation Branch of the UK. (1990). *Report on the accident to Boeing 737–400, G-OBME, near Kegworth, Leicestershire on 8 January 1989* (No. 4/1990). London: HMSO.

Aldrich, H.E. (1979). *Organizations and environments.* Englewood Cliffs, NJ: Prentice-Hall.

Aldrich, H.E. (1999). *Organizations evolving.* London: Sage Publications.

Allport, G.W. (1985). The historical background of social psychology. In G. Lindzey and E. Aronson (eds), *The handbook of social psychology* (3rd edn, vol. 1, pp. 1–46). New York: McGraw Hill.

Amason, A.C., Hochwarter, W.A., Thompson, K.R. and Harrison, A.W. (1995). Conflict: An important dimension in successful management teams. *Organizational Dynamics, 24*(2), 20–35.

American Psychological Association [APA]. (2001). *Publication manual of the American Psychological Association* (5th edn). Washington, DC: Author.

Anderson, P. (2003). *Cracking the code: The relevance of the ISM Code and its impact on shipping practices.* London: The Nautical Institute.

Anderson, P. (2008). Spelling out the need to avoid a prescriptive ISM [Interview granted The Swedish Club]. *The Swedish Club Letter, 3,* 14–15.

Argote, L., McEvily, B. and Reagans, R. (2003). Managing knowledge in organizations: An integrative framework and review of emerging themes. *Management Science, 49*(4), 571–82.

Argyris, C. (1990). *Integrating the individual and the organization.* New Brunswick, NJ: Transaction Publishers.

Argyris, C. and Schön, D.A. (1996). *Organizational learning II: Theory, method, and practice.* Reading, MA: Addison-Wesley.

Ashford, S.J., Sutcliffe, K.M. and Christianson, M.K. (2009). Speaking up and speaking out: The leadership dynamics of voice in organizations. In J. Greenberg and M.S. Edwards (eds), *Voice and silence in organizations* (pp. 175–201). Bingley, UK: Emerald Group.

Bailey, N., Ellis, N. and Sampson, H. (2006). *Perceptions of risk in the maritime industry: Ship casualty.* Cardiff: Seafarers International Research Centre (SIRC).

Baker, G.R., King, H., MacDonald, J.L. and Horbar, J.D. (2003). Using organizational assessment surveys for improvement in neonatal intensive care. *Pedriatrics, 111*(4).

Bandura, A. (1977). *Social learning theory*. Englewood Cliffs, NJ: Prentice Hall.

Bandura, A. (1997). *Self-efficacy: The exercise of control*. New York: Freeman.

Barnett, M.L., Gatfield, D.I. and Pekcan, C.H. (2003). A research agenda in maritime Crew Resource Management. In *Proceedings of the International Conference on Team Resource Management in the 21st Century*. Daytona Beach, Florida: Embry-Riddle Aeronautical University.

Barnett, M.L., Stevenson, C.J. and Lang, D.W. (2005). Shipboard manning: Alternative structures for the future? *WMU Journal of Maritime Affairs, 4*(1), 5–32.

Baskerville, R.F. (2003). Hofstede never studied culture. *Accounting, Organizations and Society, 28*(1), 1–14.

Beck, U. (1992). *Risk society: Towards a new modernity* (M. Ritter, Trans.). London: Sage.

Beck, U. (2009). *World at risk* (C. Cronin, Trans.). Cambridge: Polity Press.

Becker, H.S. (1996). The epistemology of qualitative research. In R. Jessor, A. Colby and R. Schweder (eds), *Essays on ethnography and human development*. Chicago: University of Chicago Press.

Berends, H., Kees, B. and Weggeman, M. (2003). The structuration of organizational learning. *Human Relations, 56*(9), 1035–56.

Berg, B.L. (2001). *Qualitative research methods for the social sciences*. London: Allyn & Bacon.

Bernstein, P.L. (1998). *Against the gods: The remarkable story of risk*. New York, NY: John Wiley & Sons.

Berthoin Antal, A., Dierkes, M., Child, J. and Nonaka, I. (2001a). Challenges for the future. In A. Berthoin Antal, M. Dierkes, J. Child and I. Nonaka (eds), *Handbook of organizational learning and knowledge* (pp. 921–38). New York: Oxford University Press.

Berthoin Antal, A., Dierkes, M., Child, J. and Nonaka, I. (2001b). Introduction: Factors and conditions shaping organizational learning. In A. Berthoin Antal, M. Dierkes, J. Child and I. Nonaka (eds), *Handbook of organizational learning and knowledge* (pp. 305–7). New York: Oxford University Press.

Berthoin Antal, A. and Friedman, V.J. (2005). Negotiating reality: A theory of action approach to intercultural competence. *Management Learning, 36*(1), 69–86.

Berthoin Antal, A., Krebsbach-Gnath, C. and Dierkes, M. (2003). *Hoechst challenges received wisdom on organizational learning* (Discussion paper No. SP III 2003–102). Berlin: Wissenschaftszentrum Berlin für Sozialforschung (WZB).

Bird Jr., F.E. and Loftus, R.G. (1976). *Loss control management*. Loganville, GA: Institute Press.

Bishop blasts poaching culture. (2007, 22 March). *Fairplay International Shipping News*. Retrieved 23 July 2009, from http://www.fairplay.co.uk/login. aspx?reason=denied_emptyandscript_name=/secure/display.aspxandpath_ info=/secure/display.aspxandarticlename=dn0020070322000001andphrase=B ishop.

Blau, P. (1964). *Exchange and power in social life*. New York: Wiley.

Bloor, G. and Dawson, P. (1994). Understanding professional culture in organizational context. *Organization Studies, 15*(2), 275–95.

Botterill, L. and Mazur, N. (2004). *Risk and risk perception: A literature review*. Kingston: Rural Industries Research and Development Corporation (Australian Government).

Box, G.E.P. (1979). Robustness in the strategy of scientific model building. In R.L. Launer and G.N. Wilkinson (eds), *Robustness in statistics* (pp. 201–36). New York: Academic Press.

Bresman, P.H.M. (2005). *Learning strategies and performance in organizational teams*. Unpublished PhD thesis, Massachusetts Institute of Technology, Cambridge, MA.

Brinsfield, C.T., Edwards, M.S. and Greenberg, J. (2009). Voice and silence in organizations: Historical review and current conceptualizations. In J. Greenberg and M.S. Edwards (eds), *Voice and silence in organizations* (pp. 3–33). Bingley, UK: Emerald Group.

Budworth, N. (1996). Indicators of performance in safety management. *The Safety & Health Practitioner, 14*(11), 23–9.

Bureau Enquêtes – Accidents/Mer [BEAmer]. (2003). *Abordage entre le chalutier Français Cistude et le navire-citerne Norvegien Bow Eagle: Rapport d'enquête technique*: Inspection Generale Des Service Des Affairs Maritime – BEAmer.

Cameron, I., Hare, B., Duff, R. and Maloney, B. (2006). *An investigation of approaches to worker engagement*. Norwich: HSE Books.

Campbell, S. (2006). Risk and the subjectivity of preference. *Journal of Risk Research, 9*(3), 225–42.

Carr, J.Z., Schmidt, A.M., Ford, K.J. and DeShon, R.P. (2003). Climate perceptions matter: A meta-analytic path analysis relating molar climate, cognitive and affective states, and individual level work outcomes. *Journal of Applied Psychology, 88*(4), 605–19.

Carroll, J.S. (2004). Knowledge management in high-hazard industries. In J.R. Phimister, V.M. Bier and H.C. Kunreuther (eds), *Accident precursor analysis and management: Reducing technological risk through diligence* (pp. 127–36). Washington, DC: National Academies.

Chew, W.B., Leonard-Barton, D. and Bohn, R.E. (1991). Beating Murphy's Law. *Sloan Management Review, 32*(3), 5–16.

Child, J. and Heavens, S.J. (2001). The social constitution of organizations and implications for organizational learning. In M. Dierkes, A.B. Antal, J. Child and I. Nonaka (eds), *Handbook of organizational learning and knowledge*. New York: Oxford University Press.

Cook, K.S. (2000). Advances in the microfoundations of sociology: Recent developments and new challenges for social psychology. *Contemporary Sociology, 29*(5), 685–92.

Corcoran, W.R. (2004). Defining and analyzing precursors. In J.R. Phimister, V.M. Bier and H.C. Kunreuther (eds), *Accident precursor analysis and management: Reducing technological risk through diligence* (pp. 79–88). Washington, DC: National Academies.

Cors, R. (2003). What is a learning organization? Reflections on the literature and practitioner perspectives. Retrieved 24 January 2008, from http://www.engr.wisc.edu/services/elc/lor/files/Learning_Org_Lit_Review.pdf.

Cosier, R.A. and Schwenk, C.R. (1990). Agreement and thinking alike: Ingredients for poor decisions. *Academy of Management Executive, 4*(1), 69–74.

Cox, S. and Flin, R. (1998). Safety culture: Philosopher's stone or man of straw. *Work and Stress, 12*(3), 189–201.

Crowne, D.P. and Marlowe, D. (1960). A new scale of social desirability independent of psychopathology. *Journal of Consulting Psychology, 24*(4), 349–54.

Cunningham, J.B. (1993). *Action research and organizational development.* Westport, CT: Praeger.

Dahl, S. (2004). Intercultural research: The current state of knowledge. *Middlesex University Discussion Paper No. 26*. Retrieved 25 April 2007, from http:/ssrn.com/abstract=658202.

Dalan, H. (2007). *From cost to profit: A fresh look at safety management.* Paper presented at the International Conference on Maritime Technology (ICMT) 2007, November 6, Taipei, Taiwan.

de Vaus, D. (2002). *Surveys in social research* (5th edn). London: Routledge.

Degré, T. (2008). From Black-Grey-White detention-based lists of flags to Black-Grey-White casualty-based lists of categories of vessels? *Journal of Navigation, 61*(3), 485–97.

Dekker, S. (2005). *Why we need new accident models* (Technical report No. 2005–02). Lund: Lund University School of Aviation.

Dekker, S. (2006). *The field guide to understanding human error.* Aldershot: Ashgate Publishing.

Dekker, S. and Hollnagel, E. (2004). Human factors and folk models. *Cognition, Technology and Work, 6*(2), 79–86.

Deming, W.E. (2000). *Out of the crisis* (1st MIT Press edn). Cambridge, MA: MIT Press.

Denison, D.R. (1996). What is the difference between organizational culture and organizational climate? A native's point of view on a decade of paradigm wars. *Academy of Management Review, 21*(3), 619–54.

DeVellis, R.F. (2003). *Scale development: Theory and applications* (2nd edn). Thousand Oaks, CA: Sage.

Dewberry, C. (2004). *Statistical methods for organizational research: Theory and practice.* London: Routledge.

Dewey, J. (1930). *Human nature and conduct: An introduction to social psychology.* New York: The Modern Library.

Dierkes, M., Antal, A.B., Child, J. and Nonaka, I. (eds). (2001). *Handbook of organizational learning and knowledge.* New York: Oxford University Press.

Dillman, D.A. (2007). *Mail and internet surveys: The tailored design method* (2nd edn). Hoboken, NJ: John Wiley & Sons.

Dodgson, M. (1991). Technology learning, technology strategy and competitive pressures. *British Journal of Management, 2*(3), 133–49.

Dodgson, M. (1993). Organizational learning: A review of some literatures. *Organization Studies, 14*(3), 375–94.

Dorai, R., McMurray, A.J. and Pace, R.W. (2002). *Organizational learning, change and socialization.* Paper presented at the 2002 Academy of Human Resource Development (AHRD) Conference, 27 February – 3 March, Honolulu, Hawaii.

Dörner, D. (1996). *The logic of failure: Recognizing and avoiding error in complex situations* (R. Kimber and R. Kimber, Trans.). New York: Metropolitan Books.

Druskat, V.U. and Pescosolido, A.T. (2002). The content of effective teamwork mental models in self-managing teams: Ownership, learning and heedful interrelating. *Human Relations, 55*(3), 283–314.

Dyer-Smith, M.B. (1992). Shipboard organisation: The choices for international shipping. *Journal of Navigation, 45*(3), 414–24.

Easterby-Smith, M., Araujo, L. and Burgoyne, J. (eds). (2006). *Organizational learning and the learning organization: Developments in theory and practice.* London: Sage Publications.

Easterby-Smith, M. and Lyles, M.A. (eds). (2005). *Handbook of organizational learning and knowledge management.* Malden, MA: Blackwell Publishing.

Edmondson, A.C. (1996). Learning from mistakes is easier said than done: Group and organizational influences on the detection and correction of human error. *Journal of Applied Behavioral Science, 32*(1), 5–28.

Edmondson, A.C. (1999). Psychological safety and learning behavior in work teams. *Administrative Science Quarterly, 44*(2), 350–83.

Edmondson, A.C. (2003). Managing the risks of learning: Psychological safety in work teams. In M.A. West, D. Tjosvold and K.G. Smith (eds), *International handbook of organizational teamwork and cooperative working* (pp. 255–75). Chichester, West Sussex: John Wiley & Sons.

Edmondson, A.C. (2005). Promoting experimentation for organizational learning. *Rotman Magazine, Winter,* 20–3.

Eisenberger, R., Huntington, R., Hutchison, S. and Sowa, D. (1986). Perceived organizational support. *Journal of Applied Psychology, 71*(3), 500–7.

Ek, Å. (2006). *Safety culture in sea and aviation transport.* Unpublished PhD thesis, Lund University, Lund.

Emerson, R.M. (1962). Power-dependence relations. *American Sociological Review, 27*(1), 31–41.

Emerson, R.M. (1976). Social exchange theory. *Annual Review of Sociology, 2,* 335–62.

Fearnresearch. (2009). *Fearnleys review, 2008 [includes world seaborne trade and world bulk trades, 2008]*. Oslo: Author.

Federal Railroad Administration (FRA). (2006). *Rail Crew Resource Management (CRM): Survey of teams in the railroad operating environment and identification of available CRM training methods* (No. DOT/FRA/ORD-06/10). Washington, DC: U.S. Department of Transportation.

Finger, M. and Brand, S.B. (1999). The concept of the 'learning organization' applied to the transformation of the public sector. In M. Easterby-Smith, L. Araujo and J. Burgoyne (eds), *Organizational learning and the learning organization: Developments in theory and practice*. London: Sage.

Fiol, C.M. (1994). Consensus, diversity, and learning in organizations. *Organization Science, 5*(3), 403–20.

Fiol, C.M. and Lyles, M.A. (1985). Organizational learning. *Academy of Management Review, 10*(4), 803–13.

Fischhoff, B., Lichtenstein, S., Slovic, P., Derby, S.L. and Keeney, R.L. (1981). *Acceptable risk*. New York, NY: Cambridge University Press.

Fischhoff, B., Watson, S.R. and Hope, C. (1984). Defining risk. *Policy Sciences, 17*(2), 123–39.

Flin, R., O'Connor, P. and Crichton, M. (2008). *Safety at the sharp end: A guide to non-technical skills*. Aldershot: Ashgate Publishing.

Franklin, B. (1986). *The way to wealth*. Bedford, MA: Applewood Books.

French, W.L., Kast, F.E. and Rosenzweig, J.E. (1985). *Understanding human behavior in organizations*. New York: Harper & Row.

Freudenburg, W.R. (1993). Risk and recreancy: Weber, the division of labor, and the rationality of risk perceptions. *Social Forces, 71*(4), 909–32.

Friedlander, F. (1983). Patterns of individual and organizational learning. In S. Srivastva (ed.), *The executive mind: New insights on managerial thought and action* (pp. 192–220). San Francisco, CA: Jossey-Bass.

Friedman, V.J., Lipshitz, R. and Popper, M. (2005). The mystification of organizational learning. *Journal of Management Inquiry, 14*(1), 19–30.

Galer, G.S. and van der Heijden, K. (2001). Scenarios and their contribution to organizational learning: From practice to theory. In M. Dierkes, A.B. Antal, J. Child and I. Nonaka (eds), *Handbook of organizational learning and knowledge* (pp. 849–64). New York: Oxford University Press.

García-Morales, V.J., LLoréns-Montes, F.J. and Verdú-Jover, A.J. (2006). Organisational learning categories: Their influence on organizational performance. *International Journal of Innovation and Learning, 3*(5), 518–36.

Garmston, R.J. (2005). Group wise: How to turn conflict into an effective learning process. *JSD, 26*(3), 65–6.

Garvin, D.A. (1993). Building a learning organization. *Harvard Business Review, 71*(4), 78–91.

Garvin, D.A., Edmondson, A.C. and Gino, F. (2008). Is yours a learning organization? *Harvard Business Review, 86*(3), 109–16.

Gephart, M.A., Marsick, V.J., Van Buren, M.E. and Spio, M.S. (1996). Learning organizations come alive. *Training and Development, 50*(12), 34–46.

Gibson, C.B., Zellmer-Bruhn, M.E. and Schwab, D.P. (2003). Team effectiveness in multinational organizations: Evaluation across contexts. *Group & Organization Management, 28*(4), 444–74.

Giddens, A. (1984). *The constitution of society: Outline of the Theory of Structuration*. Berkeley: University of California Press.

Giddens, A. (1990). *The consequences of modernity*. Stanford, CA: Stanford University Press.

Giddens, A. (1994). *Beyond left and right*. Stanford, CA: Stanford University Press.

Giddens, A. (ed.). (1972). *Emile Durkheim: Selected writings*. Cambridge: Cambridge University Press.

Glansdorp, C. (2009). *Final report: Maritime Operational Services (MOS)* (No. MarNIS/WP3.2/Final report/MARIN/22–12–2008/version 1.0): CETLE.

Gordon, J.R. (2002). *Organizational behavior: A diagnostic approach* (7th edn). Upper Saddle River, NJ: Prentice Hall.

Gove, P.B. (ed.). (1993). *Webster's Third New International Dictionary*. Springfield, MA: Merriam-Webster.

Graen, G.B. and Uhl-Bien, M. (1995). Relationship-based approach to leadership: Development of leader-member exchange (LMX) theory of leadership over 25 years: Applying a multi-level multi-domain perspective. *Leadership Quarterly, 6*(2), 219–47.

Graham, P. (ed.). (1995). *Mary Parker Follet: Prophet of management*. Boston: Harvard Business School Press.

Grey, M. (2009, March/April). Fatigue: Catching up with the real world. *The Sea*.

Guldenmund, F.W. (2000). The nature of safety culture: A review of theory and research. *Safety Science, 34*, 215–57.

Hale, A.R. and Glendon, A.I. (1987). *Individual behaviour in the control of danger*. Amsterdam: Elsevier.

Hall, E.T. (1976). *Beyond culture*. Garden City, NY: Anchor Press.

Hansson, S.O. (1989). Dimensions of risk. *Risk Analysis, 9*(1), 107–12.

Hansson, S.O. (2004). Philosophical perspectives on risk. *Techne, 8*(1), 10–35.

Hansson, S.O. (2007). Risk. *The Stanford Encyclopedia of Philosophy* Summer 2007 Edition. Retrieved 19 April 2007, from http://plato.stanford.edu/archives/sum2007/entries/risk/.

Harter, J.K., Schmidt, F.L. and Hayes, T.L. (2002). Business-unit-level relationship between employee satisfaction, employee engagement, and business outcomes: A meta-analysis. *Journal of Applied Psychology, 87*(2), 268–79.

Health and Safety Commission. (2006). *Improving worker involvement – Improving health and safety* (No. CD207 C10 04/06). Suffolk: Health and Safety Executive.

Hedberg, B. (1981). How organizations learn and unlearn. In P.C. Nystrom and W.H. Starbuck (eds), *Handbook of organizational design* (pp. 3–27). London: Oxford University Press.

Heinrich, H.W. (1931). *Industrial accident prevention*. New York: McGraw-Hill.

Heinrich, H.W., Petersen, D., Roos, N.R., Brown, J. and Hazlett, S. (1980). *Industrial accident prevention: A safety management approach* (5th edn). New York: McGraw-Hill.

Helmreich, R.L. (2000). On error management: Lessons from aviation. *British Medical Journal, 320,* 781–85.

Helmreich, R.L., Merritt, A.C. and Wilhelm, J.A. (1999). The evolution of Crew Resource Management training in commercial aviation. *International Journal of Aviation Psychology, 9*(1), 19–32.

Hernqvist, M. (2008, June). The importance of 'the right culture' *The Swedish Club Letter, 2,* 30–31.

Hill, T. and Lewicki, P. (2006). *Statistics: Methods and applications*. Tulsa, OK: StatSoft, Inc.

Hockey, G.R.J., Healey, A., Crawshaw, M., Wastell, D.G. and Sauer, J. (2003). Cognitive demands of collision avoidance in simulated ship control. *Human Factors, 45*(2), 252–65.

Hofmann, D.A., Morgeson, F.P. and Gerras, S.J. (2003). Climate as a moderator of the relationship between leader-member exchange and content specific citizenship: Safety climate as an exemplar. *Journal of Applied Psychology, 88*(1), 170–8.

Hofstede, G. (1997). *Cultures and organizations: Software of the mind*. New York: McGraw-Hill.

Hofstede, G. (2001). *Culture's consequences: Comparing values, behaviors, institutions and organizations across nations* (2nd edn). London: Sage Publications.

Hofstede, G. (2002). Dimensions do not exist: A reply to Brendan McSweeney. *Human Relations, 55*(11), 1355–61.

Hofstede, G. (2003). What is culture? A reply to Baskerville *Accounting, Organizations and Society, 28*(7–8), 811–13.

Hofstede, G. (2006). Dimensionalizing cultures: The Hofstede Model in context. Retrieved 10 April 2007, from http://www.ac.wwu.edu/~culture/hofstede.htm.

Holland, J.H., Holyoak, K.J., Nisbett, R.E. and Thagard, P.R. (1989). *Induction: Processes of inference, learning and discovery*. Cambridge, MA: MIT Press.

Hollnagel, E. (1998). *Cognitive reliability and error analysis method: CREAM*. Oxford: Elsevier.

Hollnagel, E. (2004). *Barriers and accident prevention*. Aldershot, England: Ashgate Publishing.

Holton, G.A. (2004). Defining risk. *Financial Analysts Journal, 60*(6), 19–25.

Homans, G.C. (1958). Social behavior as exchange. *American Journal of Sociology, 63*(6), 597–606.

House, R.J., Hanges, P.J., Javidan, M., Dorfman, P.W. and Gupta, V. (eds). (2004). *Culture, leadership, and organizations: The GLOBE study of 62 societies.* Thousand Oaks, CA: Sage Publications.

Huber, G.P. (1991). Organizational learning: The contributing processes and the literatures. *Organization Science, 2*(1), 88–115.

Inertia fatigue. (2007, 13th March). *Lloyd's List*, p. 7.

International Chamber of Shipping [ICS]. (2006). International shipping: The life blood of world trade [DVD Video]: Videotel Productions.

International Maritime Organization [IMO]. (1997). Resolution A.947(23), *Human element vision, principles and goals for the organization.* London: Author.

International Maritime Organization [IMO]. (2000). *Model course 1.21: Personal safety and social responsibility.* London: Author.

International Maritime Organization [IMO]. (2001). *International Convention on Standards of Training, Certification and Watchkeeping for Seafarers 1978, as amended in 1995 and 1997 (STCW Convention) and Seafarers' Training, Certification and Watchkeeping Code (STCW Code).* London: Author.

International Maritime Organization [IMO]. (2002a). Guidelines for Formal Safety Assessment (FSA) for use in the IMO rule-making process: MSC/Circ.1023 and MEPC/Circ.392. London: Author.

International Maritime Organization [IMO]. (2002b). *International Safety Management Code.* London: Author.

International Maritime Organization [IMO]. (2004). *Consolidated text of the International Convention for the Safety of Life at Sea, 1974, and its Protocol of 1988: Articles, annexes and certificates incorporating all amendments in effect from 1 July 2004.* London: Author.

International Maritime Organization [IMO]. (2008). *Responsibilities of Governments and measures to encourage flag state compliance: Report of the correspondence group* (No. FSI 16/3). London: International Maritime Organization.

International Organization of Standards (ISO) & International Electrotechnical Commission (IEC). (1999). *Guide 51: Safety aspects – Guidelines for their inclusion in standards* (No. ISO/IEC GUIDE 51:1999(E)). Geneva: ISO.

International Organization of Standards (ISO) & International Electrotechnical Commission (IEC). (2002). *GUIDE 73: Risk management vocabulary -Guidelines for use in standards* (No. ISO/IEC GUIDE 73:2002(E/F)). Geneva: ISO.

Jackson, N. and Carter, P. (1992). The perception of risk. In J. Ansell and F. Wharton (eds), *Risk: Analysis, assessment and management* (pp. 41–54). Chichester: John Wiley & Sons.

Janis, I.L. (1972). *Victims of groupthink.* Boston: Houghton-Mifflin.

Jenkin, C.M. (2006). Risk perception and terrorism: Applying the psychometric paradigm [Electronic Version]. *Homeland Security Affairs Journal, 2.* Retrieved 1 December 2008, from http://www.hsaj.org/?fullarticle=2.2.6.

Johnson, C.W. (2003). *Failure in safety-critical systems: A handbook of accident and incident reporting.* Glasgow, Scotland: Glasgow University Press.

Johnson, R.B. and Onwuegbuzie, A.J. (2004). Mixed methods research: A research paradigm whose time has come. *Educational Researcher, 33*(7), 14–26.

Johnson, W.G. (1980). *MORT safety assurance systems.* New York: Marcel Dekker.

Kahn, W.A. (1990). Psychological conditions of personal engagement and disengagement at work. *The Academy of Management Journal, 33*(4), 692–724.

Kanse, L. (2004). *Recovery uncovered: How people in the chemical process industry recover from failures.* Unpublished PhD dissertation, Technische Universiteit Eindhoven, Eindhoven, Netherlands.

Kanter, R.M. (1993). *Men and women of the corporation* (2nd edn). New York: Basicbooks.

Kasperson, J.X., Kasperson, R.E., Pidgeon, N. and Slovic, P. (2003). The social amplification of risk: Assessing fifteen years of research and theory. In N. Pidgeon, R.E. Kasperson and P. Slovic (eds), *The social amplification of risk* (pp. 13–46). Cambridge: Cambridge University Press.

Kasperson, R.E., Renn, O., Slovic, P., Brown, H.S., Emel, J., Goble, R., et al. (1988). The social amplification of risk: A conceptual framework. *Risk Analysis, 8*(2), 177–87.

Katz, D. and Kahn, R.L. (1978). *The social psychology of organizations* (2nd edn). New York: John Wiley and Sons.

Kelly, B.D. and Clancy, M.S. (2001). Use a comprehensive database to better manage process safety. *Chemical Engineering Progress, 97*(8), 67–9.

Kieser, A. and Koch, U. (2008). Bounded rationality and organizational learning based on rule changes. *Management Learning, 39*(3), 329–47.

Kiesler, S. and Sproull, L. (1982). Managing response to changing environments: Perspectives on problem sensing from social cognition. *Administrative Science Quarterly, 27*(4), 548–70.

Kim, D.H. (1993). *A framework and methodology for linking individual and organizational learning: Applications in TQM and product development.* Unpublished PhD thesis, Sloan School of Management: Massachusetts Institute of Technology.

Kozlowski, S.W.J. and Doherty, M.L. (1989). Integration of climate and leadership: Examination of a neglected issue. *Journal of Applied Psychology, 74*(4).

Kozlowski, S.W.J. and Ilgen, D.R. (2006). Enhancing the effectiveness of work groups and teams. *Psychological Science in the Public Interest, 7*(3), 77–124.

Krause, T.R. (2000). Motivating employees for safety success. *Professional Safety, 45*(2), 22–5.

Krause, T.R. (2005). *Leading with safety.* Hoboken, NJ: Wiley-Interscience.

Kristiansen, S. (2005). *Maritime transportation: Safety management and risk analysis.* Oxford: Elsevier Butterworth-Heinemann.

Kuo, C. (2007). *Safety management and its maritime application* London: The Nautical Institute.

Lavery, I. (1989, October). Herald of Free Enterprise legislation. *Seaways*, 15–17.

Lawler, E.J. (2001). The affect theory of social exchange. *American Journal of Sociology, 107*(2), 321–52.

Lawler, E.J. and Thye, S.R. (1999). Bringing emotions into social exchange theory. *Annual Review of Sociology, 25*, 217–44.

Lee, J.D. and Sanquist, T.F. (2000). Augmenting the operator function model with cognitive operations: Assessing the cognitive demands of technological innovation in ship navigation. *IEEE Transactions on Systems, Man and Cybernetics, Part A, 30*(3), 273–85.

Leveson, N. (2004). A new accident model for engineering safer systems. *Safety Science, 42*(4), 237–70.

Levitt, B. and March, J.G. (1988). Organizational learning. *Annual Review of Sociology, 14*, 319–40.

Lewins, A. and Silver, C. (2007). *Using software in qualitative research: A step-by-step guide.* London: Sage Publications.

Liang, D.W., Moreland, R. and Argote, L. (1995). Group versus individual training and group performance: The mediating role of transactive memory. *Personality and Social Psychology Bulletin, 21*(4), 384–93.

Lingard, L., Espin, S., Whyte, S., Regehr, G., Baker, G.R., Reznick, R., et al. (2004). Communication failures in the operating room: An observational classification of recurrent types and effects. *Quality Safety Health Care, 13*, 330–4.

Lloyd's Register-Fairplay. (2009). *World fleet statistics, 2008: Statistical summary of the current world fleet of propelled sea-going merchant ships of not less than 100 GT as at the end of year, and of those ships completed during the year.* Redhill: Author.

Luhmann, N. (2006). *Risk: A sociological theory* (R. Barrett, Trans.). New Brunswick, NJ: Aldine Transaction.

Magala, S. (2004). *Cross-cultural compromises, multiculturalism and the actuality of unzipped Hofstede* (No. ERS-2004-078-ORG). Rotterdam: Erasmus Research Institute of Management (ERIM); Rotterdam School of Management/Rotterdam School of Economics.

Maier, G.W., Prange, C. and Rosenstiel, L. v. (2001). Psychological perspectives of organizational learning. In M. Dierkes, A.B. Antal, J. Child and I. Nonaka (eds), *Handbook of organizational learning and knowledge.* New York: Oxford University Press.

Maloney, W.F. and Cameron, I. (2003). *Employee involvement, consultation and information sharing in health and safety in construction* (No. GR/S25494/01). Swindon, UK: Engineering and Physical Science Research Council.

Manuele, F.A. (2003). *On the practice of safety* (3rd edn). Hoboken, NJ: John Wiley & Sons.

Manuele, F.A. and Main, B.W. (2002). On acceptable risk: What level of risk, if any is acceptable in the workplace? *Occupational Hazards, 64*(1), 57–60.

March, J.G., Sproull, L.S. and Tamuz, M. (1991). Learning from samples of one or fewer. *Organization Science, 2*(1), 1–13.

Marcus, A.A. and Nichols, M.L. (1999). On the edge: Heeding the warnings of unusual events. *Organization Science, 10*(4), 482–99.

Marine Accident Investigation Branch. (1999). *Marine Accident Report of the Inspector's Inquiry into the loss of MV Green Lily on the 19 November 1997 off the East Coast of Bressay, Shetland Islands* (No. 5/99). London: Author.

Marsick, V.J. and Watkins, K.E. (2003). Demonstrating the value of an organization's learning culture: The Dimensions of Learning Organization Questionnaire. *Advances in Developing Human Resources, 5*(2), 132–51.

Martin, M. (2000). *Verstehen: The uses of understanding in social science.* New Brunswick, NJ: Transaction Publishers.

Marx, K. (1904). *A contribution to the critique of political economy* (N.I. Stone, Trans.). Chicago: Charles H. Kerr Publishing.

Maslow, A.H. (1966). *The psychology of science: A reconnaissance.* New York: Harper & Row.

Mathiesen, T.C. (1996). Safety management and loss control. In *BIMCO Review 1996* (pp. 133–4). London: Stroudgate Plc. for BIMCO.

May, D.R., Gilson, R.L. and Harter, L.M. (2004). The psychological conditions of meaningfulness, safety and availability and the engagement of the human spirit at work. *Journal of Occupational and Organizational Psychology, 77*, 11–37.

McGrath, J.E. and Tschan, F. (2004). Dynamics in groups and teams: Groups as complex action systems. In A.H. Van de Ven (ed.), *Handbook of organizational change and innovation* (pp. 50–72). Cary, NC: Oxford University Press.

McHugh, M.B. (2007, October). Academic is wrong on the attractions of seafaring. *Nautilus UK Telegraph, 40*, 18.

McSween, T.E. (2003). *Value-based safety process: Improving your safety culture with behavior-based safety* (2nd edn). Hoboken, NJ: Wiley-Interscience.

McSweeney, B. (2002a). The essentials of scholarship: A reply to Geert Hofstede. *Human Relations, 55*(11), 1363–72.

McSweeney, B. (2002b). Hofstede's model of national cultural differences and their consequences: A triumph of faith – a failure of analysis. *Human Relations, 55*(1), 89–118.

Meeker, B.F. (1971). Decisions and exchange. *American Sociological Review, 36*(3), 485–95.

Meijer, F. (1986). *A history of seafaring in the classical world.* London: Taylor & Francis.

Miles, M.B. and Huberman, M.A. (1994). *Qualitative data analysis: An expanded sourcebook* (2nd edn). London: Sage Publications.

Miller, D. (1990). *The Icarus paradox: How exceptional companies bring about their own downfall.* New York, NY: Harper Collins Publishers.

Miller, D. (1992). The Icarus paradox: How exceptional companies bring about their own downfall. *Business Horizons, 35*(1), 24–35.

Miller, D. (1996). A preliminary typology of organizational learning: Synthesizing the literature. *Journal of Management, 22*(3), 485–505.

Mills, A.J. (2007a). Introducing organizational behaviour. In A.J. Mills, J.C. Helms Mills, C. Forshaw and J. Bratton (eds), *Organizational behaviour in a global context*. Peterborough, Ontario: Broadview Press.

Mills, A.J. (2007b). Understanding organisational behaviour in context. In A.J. Mills, J.C. Helms Mills, C. Forshaw and J. Bratton (eds), *Organizational behaviour in a global context*. Peterborough, Ontario: Broadview Press.

Mol, T. (2003). *Productive Safety Management*. Boston, MA: Butterworth-Heinemann.

Möller, N., Hansson, S.O. and Peterson, M. (2006). Safety is more than the antonym of risk. *Journal of Applied Philosophy, 23*(4), 419–31.

Morgan, G. (2006). *Images of organization* (Updated edn). Thousand Oaks, CA: Sage Publications.

Muhr, T. and Friese, S. (2004). *User's manual for ATLAS.ti 5.0* (2nd edn with addendum for new features in ATLAS.ti 6.0 edn). Berlin: Scientific Software Development.

Mukkadayil, J.P. (2001). *Tanker accidents: Double hull is not the only viable alternative*. Unpublished MSc. dissertation, World Maritime University, Malmö.

Muller, D. and Ornstein, K. (2007). Perceptions of and attitudes towards medical errors among medical trainees. *Medical Education, 41*(7), 645–52.

Mullins, L.J. (2005). *Management and organisational behaviour* (7th edn). Essex: Prentice Education Limited.

Murray, J.A.H., Bradley, H., Craigie, W.A., Onions, C.T. and Burchfield, R.W. (eds). (1991). *Compact Oxford English Dictionary* (2nd edn). New York: Oxford University Press.

Mythen, G. (2004). *Ulrich Beck: A critical introduction to the risk society*. London: Pluto Press.

Naot, B.-H.Y., Lipshitz, R. and Popper, M. (2004). Discerning the quality of organizational learning. *Management Learning, 35*(4), 451–72.

Nathan, M.L. and Kovoor-Misra, S. (2002). No pain, yet gain: Vicarious organizational learning from crises in an inter-organizational field. *The Journal of Applied Behavioral Science, 38*(2), 245–66.

National Transportation Safety Board [NTSB]. (2007). *Marine accident brief: Boiler rapture on the S/S Norway* (Accident investigation report No. DCA-03-MM-032). Washington, DC: NTSB.

Nembhard, I.M. and Edmondson, A.C. (2006). Making it safe: The effects of leader inclusiveness and professional status on psychological safety and improvement efforts in health care teams. *Journal of Organizational Behaviour, 27*, 941–66.

Nicolini, D. and Meznar, M.B. (1995). The social construction of organizational learning: Conceptual and practical issues in the field. *Human Relations, 48*(7), 727–46.

O'Neil, W.A. (2000). Quality shipping and the role of IMO. Retrieved 15 July 2009, from http://www.imo.org/About/mainframe.asp?topic_id=83anddoc_id=375 Officer of Norwegian tanker jailed over collision. (2003). *Nordic Business Report*. Retrieved 24 June 2008, from http://findarticles.com/p/articles/mi_m0HXI/is_2003_March_14/ai_n25064331?tag=artBody;col1.

Oil Companies International Marine Forum [OCIMF]. (2008). *Tanker management and self assessment 2: A best practice guide for vessel operators* (2nd edn). London: Author.

Oppenheim, A.L. (1954). The seafaring merchants of Ur. *Journal of the American Oriental Society, 74*(1), 6–17.

Osland, J.S. and Bird, A. (2000). Beyond sophisticated stereotyping: Cultural sensemaking in context. *Academy of Management Executive, 14*(1), 65–77.

Otway, H. (1992). Public wisdom, expert fallibility: Toward a contextual theory of risk. In S. Krimsky and D. Golding (eds), *Social theories of risk* (pp. 215–28). Westport, CT: Praeger.

Pace, R.W. (2002). The organizational learning audit. *Management Communication Quarterly, 15*(3), 458–65.

Pallant, J. (2007). *SPSS survival manual* (3rd edn). Berkshire, England: McGraw-Hill.

Patton, M.Q. (2002). *Qualitative evaluation and research methods* (3rd edn). Thousand Oaks, CA: Sage.

Pawlowsky, P. (2001). The treatment of organizational learning in management science. In M. Dierkes, A.B. Antal, J. Child and I. Nonaka (eds), *Handbook of organizational learning and knowledge*. New York: Oxford University Press.

Pearsall, J. (ed.). (2001). *New Oxford Dictionary*. New York: Oxford University Press.

Perrow, C. (1999). *Normal accidents: Living with high-risk technologies*. Princeton, NJ: Princeton University Press.

Peters, T. (1978). Symbols, patterns and settings: An optimistic case for getting things done. *Organizational Dynamics, 7*, 3–23.

Pettigrew, A.M. (1979). On studying organizational culture. *Administrative Science Quarterly, 24*(4), 570–81.

Pidgeon, N. (1997). The limits to safety? Culture, politics, learning and man-made disasters. *Journal of Contingencies and Crisis Management, 5*(1), 1–14.

Pidgeon, N. and O'Leary, M. (2000). Man-made disasters: Why technology and organizations (sometimes) fail. *Safety Science, 34*(1–3), 15–30.

Porter, L.W. and Lawler, E.E. (1968). *Managerial attitudes and performance*. Homewood, IL: R.D. Irwin.

Precious Associates Limited and D.M. Jupe Consulting. (2008). *Drewry annual report: Manning – 2008*. London: Drewry Shipping Consultants.

Quinton, A. (1995). Philosophy. In T. Honderich (ed.), *The Oxford companion to philosophy* (pp. 666–70). New York: Oxford University Press.

Qureshi, Z.H. (2007). *A review of accident modelling approaches for complex sociotechnical systems.* Paper presented at the 12th Australian Workshop on Safety Related Programmable Systems (SCS '07), Adelaide, Australia.

Qureshi, Z.H. (2008). *A review of accident modelling approaches for complex critical sociotechnical systems* (No. DSTO-TR-2094). Edinburgh, Australia: Defence Science and Technology Organisation of the Australian Government.

Rasmussen, J. (1997). Risk management in a dynamic society: A modelling problem. *Safety Science, 27*(2/3), 183–213.

Rasmussen, J. and Svedung, I. (2000). *Proactive risk management in a dynamic society.* Borås: Swedish Rescue Services Agency.

Reason, J. (1990). *Human error.* Cambridge: Cambridge University Press.

Reason, J. (1997). *Managing the risks of organizational accidents.* Aldershot: Ashgate Publishing.

Redding, J. (1997). Hardwiring the learning organization. *Training and Development, 51*(8), 61–7.

Reichers, A.E. and Schneider, B. (1990). Climate and culture: An evolution of constructs. In B. Schneider (ed.), *Organizational climate and culture* (pp. 5–39). San Fransico, CA: Jossey-Bass.

Renn, O. (1992). Concepts of risk: A classification. In S. Krimsky and D. Golding (eds), *Social theories of risk* (pp. 53–79). Westport, CT: Praeger.

Renn, O. (2005). *Risk governance: Towards an integrative approach* (IRGC white paper no. 1). Geneva: International Risk Governance Council.

Report on the loss of the S.S. *Titanic.* (1990). Gloucester: Alan Sutton Publishing.

Rhoades, L. and Eisenberger, R. (2002). Perceived organizational support: A review of the literature. *Journal of Applied Psychology, 87*(4), 698–714.

Richardsen, P.W. (2008). Expensive safety hangover in the shipping industry. Retrieved 8 August 2008, from ttp://www.dnv.com/press_area/press_ releases/2008/expensivesafetyhangoverintheshippingindustry.asp.

Robertson, V. and Stewart, T. (2004). *Risk perception in relation to musculoskeletal disorders.* Norwich: HSE Books.

Rodan, S. (2005). Exploration and exploitation revisited: Extending March's model of mutual learning. *Scandinavian Journal of Management, 21*(4), 407–28.

Rosa, E.A. (2003). The logical structure of the social amplification of risk (SARF): Metatheoretical foundations and policy implications. In N. Pidgeon, R.E. Kasperson and P. Slovic (eds), *The social amplification of risk* (pp. 47–79). Cambridge: Cambridge University Press.

Rosenheck, R.A. (2001). Organizational process: A missing link between research and practice. *Psychiatric Services, 52*(12), 1607–12.

Rosness, R., Guttormsen, G., Steiro, T., Tinmannsvik, R.K. and Herrera, I.A. (2004). *Organisational accidents and resilient organisations: Five perspectives (Revision 1)* (No. STF38 A 04403). Trondheim: SINTEF Industrial Management.

Rossman, G.B. and Wilson, B.L. (1985). Numbers and words: Combining quantitative and qualitative methods in a single large-scale evaluation study. *Evaluation Review, 9*(5), 627–43.

Rossman, G.B. and Wilson, B.L. (1991). *Numbers and words revisited: Being 'shamelessly eclectic.'* Paper presented at the Annual Meeting of the American Educational Research Association (3–7April), Chicago, IL.

Rothman, J. and Friedman, V.J. (2001). Identity, conflict and organizational learning. In M. Dierkes, A.B. Antal, J. Child and I. Nonaka (eds), *Handbook of organizational learning and knowledge*. New York: Oxford University Press.

Runciman, W.G. (ed.). (1978). *Max Weber: Selections in translation*. Cambridge: Cambridge University Press.

Rundmo, T. (1992). Risk perception and safety on offshore petroleum platforms – Part I: Perception of risk. *Safety Science, 15*, 39–52.

Ryan, R.M. and Deci, E.L. (2000). Intrinsic and extrinsic motivations: classic definitions and new directions. *Contemporary Educational Psychology, 25*, 54–67.

Sadler, P. (2001). Leadership and organizational learning. In M. Dierkes, A.B. Antal, J. Child and I. Nonaka (eds), *Handbook of organizational learning and knowledge*. New York: Oxford University Press.

Sagan, S.D. (1993). *The limits of safety: Organizations, accidents, and nuclear weapons*. Princeton: Princeton University Press.

Sagan, S.D. (1994). Toward a political theory of organizational reliability. *Journal of Contingencies and Crisis Management, 2*(4), 228–240.

Sampson, H. (2002, Nov-Dec). Destructive obedience and the importance of seafarer training. *The Sea*, p. 4.

Schein, E.H. (2004). *Organizational culture and leadership* (3rd edn). San Francisco: Jossey-Bass.

Schuck, P.H. (2000). *The limits of law*. Boulder, Colorado: Westview.

Schwartz, S.H. (1994). Beyond individualism/collectivism: New cultural dimensions of values. In U. Kim, H.C. Triandis, S.-C. Kagitcibasi and G. Yoon (eds), *Individualism and collectivism: Theory, methods and applications* (pp. 85–119). London: Sage.

Senge, P.M. (2006). *The fifth discipline: The art and practice of the learning organization* (Revised edn). New York: Doubleday/Currency Books.

Shapiro, S.S. and Wilk, M.B. (1965). An analysis of variance test for normality (complete samples). *Biometrika, 52*(3/4), 591–611.

Shearn, P. (2004). *Workforce participation in the management of occupational health and safety*. Sheffield: Health and Safety Laboratory.

Silverman, D. (2005). *Doing qualitative research* (2nd edn). London: Sage Publications.

Simon, H.A. (1982). *Models of bounded rationality* (vols 1–3). Cambridge, MA: MIT Press.

Simon, H.A. (1991). Bounded rationality and organizational learning. *Organization Science, 2*(1), 125–34.

Singleton Jr., R.A. and Straits, B.C. (2005). *Approaches to social research.* Oxford: Oxford University Press.

Sjöberg, L. (2000). The methodology of risk perception research. *Quality and Quantity, 34*(4), 407–18.

Sjöberg, L. (2006). Rational risk perception: Utopia or dystopia? *Journal of Risk Research, 9*(6), 683–96.

Skjong, R. (2005). Risk: Not quit Greek to DNV. In E. Halvorsen (ed.), *DNV Forum: Surveying with class.* Høvik: DNV Corporate Communications.

Slovic, P. (1992). Perception of risk. Reflections on the psychometric paradigm. In S. Krimsky and D. Golding (eds), *Social theories of risk* (pp. 117–52). Westport, CT: Praeger.

Slovic, P. (1999). Trust, emotion, sex, politics, and science: Surveying the risk assessment battlefield. *Risk Analysis, 19*(4), 689–701.

Slovic, P. and Weber, E.U. (2002). *Perception of risk posed by extreme events.* Paper presented at the Conference on Risk Management Strategies in an Uncertain World (12–13 April), Palisades, New York.

Smircich, L. (1983). Concepts of culture and organizational analysis. *Administrative Science Quarterly, 28*(3), 339–58.

Smith, A. (1976). *An inquiry into the nature and causes of the wealth of nations (Book V)* (Cannan's edn). Chicago: University of Chicago Press.

Spencer-Oatey, H. (2000). *Culturally speaking: Managing rapport through talk across cultures.* London: Continuum.

Starbuck, W.H. and Hedberg, B. (2001). How organizations learn from success and failure. In M. Dierkes, A.B. Antal, J. Child and I. Nonaka (eds), *Handbook of organizational learning and knowledge.* New York: Oxford University Press.

Stata, R. (1989). Organizational learning: The key to management innovation. *Sloan Management Review, 30*(3), 63–74.

Strahan, R. and Gerbasi, K.C. (1972). Short, homogeneous versions of the Marlowe-Crowne Social Desirability Scale. *Journal of Clinical Psychology, 28*, 191–93.

Sudhakar, U.R.P. (2005). *Promoting safety culture in shipping: issues and strategies.* Paper presented at the Institute of Marine Engineers (India) – Visakhapatnam Branch, 15 April 2005, Visakhapatnam, India.

Susarla, A. (2003). Plague and arsenic: Assignment of blame in the mass media and the social amplification and attenuation of risk. In N. Pidgeon, R.E. Kasperson and P. Slovic (eds), *The social amplification of risk* (pp. 179–206). Cambridge: Cambridge University Press.

Svedung, I. and Rådbo, H. (2006). Feedback for pro-activity: Who should learn what from events, when and how. In O. Svenson, I. Salo, A.B. Skjerve, T. Reiman and P. Oedewald (eds), *Nordic perspectives on safety management in high reliability organizations: Theory and applications* (pp. 21–34). Stockholm: Stockholm University, Dept. of Psychology.

Swidler, A. (1986). Culture in action: Symbols and strategies. *American Sociological Review, 51*(2), 273–86.

Tabachnick, B.G. and Fidell, L.S. (2007). *Using multivariate statistics* (5th edn). Boston, MA: Pearson Education.

Taleb, N.N. (2007). *The Black Swan: The impact of the highly improbable.* London: Penguin Books.

Tamuz, M. (2004). Understanding accident precursors. In J.R. Phimister, V.M. Bier and H.C. Kunreuther (eds), *Accident precursor analysis and management: Reducing technological risk through diligence.* Washington, DC: National Academies.

Teichmann, J. and Evans, K.C. (1999). *Philosophy: A beginner's guide* (3rd edn). Oxford: Blackwell Publishing.

Templeton, G.F., Morris, S.A., Snyder, C.A. and Lewis, B.R. (2004). Methodological and thematic prescriptions for defining and measuring the organizational learning concept. *Information System Frontiers, 6*(3), 263–76.

Thai, V.V. and Grewal, D. (2006). The Maritime Safety Management System (MSMS): A survey of the international shipping community. *Maritime Economics and Logistics, 8*(3), 287–310.

Thibaut, J. and Kelley, H.H. (1959). *The social psychology of groups.* New York: Wiley.

Thompson, P.B. (1990). Risk objectivism and risk subjectivism: When are risks real? *Risk: Issues in Health and Safety, 1,* 3–22.

Thompson, P.B. (1992). Reply to Valverde. *Risk: Issues in Health & Safety, 3,* 49 ff.

Trochim, W. and Donnelly, J.P. (2007). *The research methods knowledge base* (3rd edn). Mason, OH: Thomson.

Trompenaars, F. and Hampden-Turner, C. (1997). *Riding the waves of culture: Understanding cultural diversity in business* (2nd edn). London: Brealey.

Turner, B.A. and Pidgeon, N.F. (1997). *Man-made disasters* (2nd edn). Oxford: Butterworth-Heinemann.

Tynan, R. (2005). The effects of threat sensitivity and face giving on dyadic psychological safety and upward communication. *Journal of Applied Social Psychology, 35*(2), 223–47.

Tyre, M.J. and von Hippel, E. (1997). The situated nature of adaptive learning in organizations. *Organization Science, 8*(1), 71–83.

UK Department of Transport. (1987). *MV Herald of Free Enterprise (formal investigation): Report of court number 8074.* London: Her Majesty's Stationery Office.

United States Coast Guard [USCG]. (2005). *Investigation into the explosion and sinking of the chemical tanker Bow Mariner* (No. 16732). Washington, DC: US Department of Homeland Security.

Vago, S. (2006). *Law and society* (8th edn). Upper Saddle River, NJ: Pearson Prentice Hall.

Valverde, L.J.A., Jr. (1991). The cognitive status of risk: A response to Thompson. *Risk: Issues in Health & Safety, 2,* 313 ff.

van der Schaaf, T.W., Frese, M. and Heimbeck, D. (1996). Human recovery and error management. In H.G. Stassen (ed.), *Proceedings of the XVth Annual Conference on Human Decision Making and Manual Control*. Delft: Technische Universiteit.

Vaughan, D. (1996). *The Challenger launch decision: Risky technology, culture, and deviance at NASA*. Chicago: University of Chicago Press.

Vaughan, D. (1999a). The dark side of organizations: Mistake, misconduct, and disaster. *Annual Review of Sociology, 25*, 271–305.

Vaughan, D. (1999b). The role of the organization in the production of technoscientific knowledge. *Social Studies of Science, 29*(6), 913–43.

Vroom, V.H. (1964). *Work and motivation*. New York: John Wiley & Sons.

Wagenaar, W.A. and Groeneweg, J. (1987). Accidents at sea: Multiple causes and impossible consequences. *International Journal of Man-Machine Studies, 27*(5/6), 587–98.

Wang, C.L. and Ahmed, P.K. (2002). *A review of the concept of organizational learning* (No. WP004/02). Telford, Shropshire, UK: Wolverhampton Business School, University of Wolverhampton.

Waring, A.E. and Glendon, A.I. (1998). *Managing risk*. London: International Thomson Business Press.

Warwick Institute for Employment Research. (2005). *BIMCO/ISF manpower 2005 update: The worldwide demand for and supply of seafarers – main report*. Coventry, UK: BIMCO/ISF.

Wegner, D.M. (1986). Transactive memory: A contemporary analysis of the group mind. In G. Mullen and G. Goethals (eds), *Theories of group behavior* (pp. 185–208). New York: Springer-Verlag.

Westrum, R. (1993). Cultures with requisite imagination. In J.A. Wise, D.V. Hopkin and P. Stager (eds), *Verification and validation of complex systems: Human factors issues* (pp. 401–16). Berlin: Springer-Verlag.

Wheatley, M.J. (1999). *Leadership and the new science: Discovering order in a chaotic world* (2nd edn). San Francisco, CA: Berrett-Koehler Publishers.

Williamson, D. (2002). Forward from a critique of Hofstede's model of national culture. *Human Relations, 55*(11), 1373–95.

Wilson, E. and Rees, C. (2007). Perception, stereotyping and attribution. In A.J. Mills, J.C. Helms Mills, C. Forshaw and J. Bratton (eds), *Organizational behavior in a global context*. Peterborough, Ontario: Broadview Press.

Wilson, F.M. (2004). *Organizational behaviour and work: A critical introduction* (2nd edn). Oxford: Oxford University Press.

Woods, D. and Cook, R. (2001). From counting failures to anticipating risks: Possible futures for patient safety. In L. Zipperer and S. Cushman (eds), *Lessons in patient safety: A primer* (pp. 89–97). Chicago, IL: National Patient Safety Foundation.

Yamagishi, T., Gillmore, M.R. and Cook, K.S. (1988). Network connections and the distribution of power in exchange networks. *The American Journal of Sociology, 93*(4), 833–51.

Zachcial, M. (ed.). (2008). *Shipping statistics yearbook, 2008*. Bremen: Institute of Shipping Economics and Logistics (ISL).

Zafirovski, M. (2005). Social Exchange Theory under scrutiny: A positive critique of its economic-behaviorist formulations [Electronic Version]. *Electronic Journal of Sociology*. Retrieved 7 July 2008, from http://www.sociology.org/content/2005/tier2/SETheory.pdf.

Zuber-Skerritt, O. (ed.). (1996). *New directions in action research*. London: Falmer Press.

Index